AF564715

# Water and Soil: Chemistry, Risk and Management

# Water and Soil: Chemistry, Risk and Management

Prof. Sriniwas Puri

RANDOM PUBLICATIONS
NEW DELHI (INDIA)

**Water and Soil: Chemistry, Risk and Management**

ISBN 978-93-5111-596-0


Published in 2015 in India by

**RANDOM PUBLICATIONS**

4376-A/4B, Gali Murari Lal, Ansari Road
New Delhi-110 002
Phone : +9111-43580356, 011-23289044, 011-43142548
e-mail: sales@randompublications.com,
info@randompublications.com, randomexports@gmail.com

Reprinted 2025

*Type Setting by* : Friends Media, Delhi-110089
*Digitally Printed at*: Replika Press Pvt. Ltd.

# Preface

Soils can process and contain considerable amounts of water. They can take in water, and will keep doing so until they are full, or until the rate at which they can transmit water into and through the pores is exceeded. Some of this water will steadily drain through the soil and end up in the waterways and streams, but much of it will be retained, away from the influence of gravity, for use of plants and other organisms to contribute to land productivity and soil health. Pores provide for the passage and/or retention of gasses and moisture within the soil profile. The soil's ability to retain water is strongly related to particle size; water molecules hold more tightly to the fine particles of a clay soil than to coarser particles of a sandy soil, so clays generally retain more water. Conversely, sands provide easier passage or transmission of water through the profile. Clay type, organic content, and soil structure also influence soil water retention. The maximum amount of water that a given soil can retain is called field capacity, whereas a soil so dry that plants cannot liberate the remaining moisture from the soil particles is said to be at wilting point. Available water is that which the plants can utilize from the soil within the range of field capacity and wilting point. The role of soil water retention is profound; its effects are far reaching and relationships are invariably complex. This section focuses on a few key roles and recognizes that it is beyond the scope of this discussion to encompass all roles that can be found in the literature.

Water and soil: chemistry, risk and management retention is widely studied and reported on in the literature due to its extensive and profound role; it is a key part of the hydrological cycle, provides support to organisms, interacts with climate, and is a major consideration in ground and surface water supply, environmental and geo-technical aspects.

I would like to thank my team for standing beside me throughout my career and writing this book. My special thanks go to "RandomPublications" who have published the book.

*– Prof. Sriniwas Puri*

# Contents

# 1

# Water Chemistry

Water chemistry plays an important role in the health, abundance and diversity of the aquatic life that can live in a stream. Excessive amounts of some constituents, such as nutrients, or the lack of others, such as dissolved oxygen, can result in degraded conditions and harm aquatic life.

## TEMPERATURE

Water temperature is important because most of the physical, chemical, and biological characteristics of a river are directly affected by temperature

### TEMPERATURE CAN AFFECT THE FOLLOWING FACTORS

1. The amount of gas, including oxygen, that can be dissolved in the water; cold water can hold more oxygen than warm water
2. The rate of photosynthesis by algae and other aquatic plants
3. The metabolic rates of aquatic organisms (increased respiration, digestion, etc.)
4. Organisms can become more sensitive. Increased metabolic rates result in the organism being stressed and more vulnerable to disease, parasites, and pollution.

### WHAT IMPACTS STREAM TEMPERATURE?

1. Air temperature. The natural seasonal changes in temperature impact stream temperature.
2. Thermal pollution occurs when water entering the stream is warmer than the water already present in the stream. One source is from industries like nuclear power plants, which discharge cooling water. Another source is stormwater run-off from heated surfaces such as parking lots, streets, roofs, etc.
3. Riparian cover removal. The removal of trees impacts water temperature by eliminating shade along the river and allowing more soil particles to reach the stream.
4. Soil erosion increases the amount of suspended solids carried by the

water. Particles in cloudy water absorb radiation from the sun, which warms the water.

## DISSOLVED OXYGEN

### WHAT IS DISSOLVED OXYGEN?

Dissolved oxygen (DO) is essential to the survival of organisms in a stream. The presence of oxygen is a positive sign and the absence of oxygen is a sign of severe pollution. Waters with consistently high dissolved oxygen are considered to be stable aquatic systems capable of supporting diverse aquatic life.

### SOURCES OF DISSOLVED OXYGEN

1. Atmosphere. The air we breathe contains approximately 21 per cent oxygen which equates to 210,000 ppm oxygen. Most surface waters contain between 5 and 15 ppm DO.
2. Photosynthesis by algae and rooted aquatic plants. Plants deliver oxygen to water through photosynthesis.

### NATURAL INFLUENCES ON DISSOLVED OXYGEN

1. Seasonal Temperature Changes. Gases, like oxygen, are more easily dissolved in cooler water than in warmer water. DO levels may be higher in winter than in summer.
2. Stream Discharge. Dry periods often result in severely reduced stream discharge and increased water temperatures. This combination acts to reduce DO levels. Wet weather or melting snows increase stream discharge and the possibility for mixing of atmospheric oxygen.
3. Dissolved or suspended solids. Oxygen dissolves more readily in water that does not contain a high concentration of salts, minerals, or other solids.
4. Aquatic Plants.
   a. Density. The density of aquatic plants will affect DO. Fewer plants grow in winter because of cold temperature and shorter day length. More green plants mean more photosynthesis, which produces more oxygen during the day when the sun is shining
   b. Respiration. Conversely, plants and animals respire 24 hours per day, using oxygen and producing carbon dioxide. So there is a net gain in DO as long as photosynthesis is occurring. In addition, when plants die, DO is used in the decomposition process. This occurs when organic matter decomposes. If there is an overabundance of algae or other aquatic plants and a large die-off occurs at one time, DO can be dramatically impacted.

## DIEL OXYGEN FLUCTUATION

The term, "diel" refers to a 24-hour period that usually includes a day and adjoining night. During the daylight hours, DO levels rise due to plant photosynthesis. When the sun sets, photosynthesis stops and respiration continues. As a result, DO levels naturally drop overnight, reaching their lowest level just before dawn, at which time the sun rises and photosynthesis again pumps more DO into the water. Extensive algal growth can result in large fluctuations in oxygen from late afternoon to early morning. If the DO levels fall too low, aquatic animals can die. This is more problematic in ponds and in backwaters of streams where flow is non-existent or very slow.

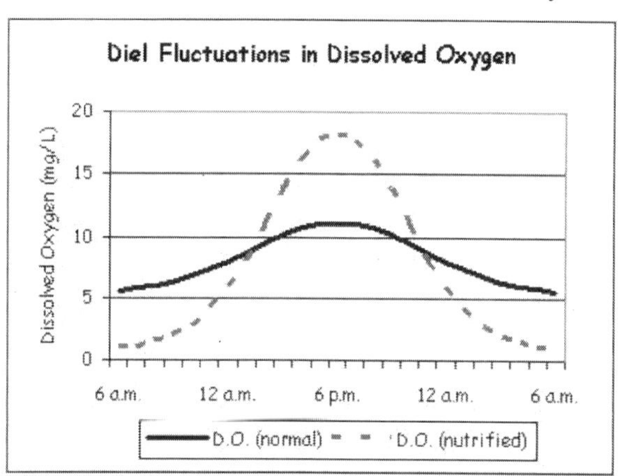


## HUMAN-CAUSED CHANGES IN DISSOLVED OXYGEN

1. Organic waste. This includes waste from once-living plants and animals and from animal feces. Excessive organic waste often comes from sewage treatment plants, malfunctioning septic systems, or manure run-off from animal operations. Organic waste can act as a fertilizer to stimulate aquatic plant growth. In time, these plants die, and they too become organic waste.
2. Urban run-off. Rain carries heat, salt, sediment, and other pollutants from impervious surfaces (streets, roofs, parking lots, etc.) into streams. This raises the water temperatures and total solids in the water reducing its capacity to hold DO.
3. Dams. Some dams are constructed so water is released from the bottom of a lake or reservoir. Seasonally this water can be almost devoid of oxygen. The opposite problem can occur when water is released from the top of a dam or spillway. This can cause excessive uptake of air from the atmosphere and result in water that has too much atmospheric gas.
4. Removal of vegetation, especially trees, in the riparian corridor. A

lack of shade causes increased water temperature and a lack of protection from erosion. This causes increased suspended solids that can work together to reduce oxygen levels.

## NITRATES AND AMMONIA

### WHAT IS NITROGEN?

Nitrogen is required by all living plants and animals for building protein. In aquatic ecosystems, nitrogen is present in different forms. The usable forms of nitrogen for aquatic plant growth are ammonia ($NH_3$) and nitrate ($NO_3$). Excess amounts of nitrogen compounds can result in unusually large populations of aquatic plants and/or organisms that feed on plants. For instance, algal blooms can be a result of excess nitrogen. As aquatic plants and animals die, bacteria break down the organic matter.

Ammonia ($NH_3$ or $NH_4$) is oxidized, or combined with oxygen ($O_2$), to form nitrites ($NO_2$) and nitrates ($NO_3$)

$$NH_3 + O_2 \rightarrow NO_2 + O_2 \rightarrow NO_3$$

The cycle for breaking down organic matter (both the biological process and the chemical process) uses up dissolved oxygen. Sources of Excess Nitrates and Ammonia in Streams:

1. Poorly functioning septic systems
2. Inadequately treated wastewater from sewage treatment plants
3. Storm drains
4. Run-off from feed lots
5. Run-off from crop fields, parks, and lawns

## PHOSPHORUS

### WHAT IS PHOSPHORUS?

Phosphorus usually takes the form of phosphate (PO4) in water. Phosphorus is also a plant nutrient and is often the limiting nutrient for plant growth, as it is less prevalent in surface water than nitrogen. Small increases in phosphorus, however, can result in a large impact on the growth of aquatic plants.

Phosphorus binds readily with soil particles. Soil must be highly saturated with phosphorus before excess amounts are detectable in shallow groundwater that can enter streams and cause negative impacts.

### SOURCES OF EXCESS PHOSPHORUS IN STREAMS

1. Septic systems and wastewater from sewage treatment plants
2. Run-off from feed lots and from the application of animal wastes on fields

3. Run-off of commercial fertilizer from crop fields, lawns, golf courses, or parks

## TURBIDITY

### WHAT IS TURBIDITY?

Turbidity testing measures the clarity of water. Low turbidity water is clear while high turbidity water is cloudy or murky. Cloudy water is most often caused by suspended matter (such as soil particles) and plankton (such as diatoms). By evaluating how turbid the water is, you can evaluate whether excess soil erosion or algal growth is occurring.

### SOURCES OF EXCESS TURBIDITY IN STREAMS

1. Development, construction, and land disturbance can result in soil erosion.
2. Quarries and gravel mining operations can result in fine sediment entering a stream and smothering habitat.
3. Agricultural areas that have not adopted "best management practices" to prevent soil erosion leak sediment to local waterways.

## pH (PARTS HYDROGEN)

### WHAT IS pH?

Water ($H_2O$) contains both $H^+$ (hydrogen) ions and $OH^-$ (hydroxide) ions. "pH" is an abbreviation for the French expression, "Pouvoir Hydrogene," meaning "the power of Hydrogen." It measures the $H^+$ ion concentration of substances and gives the results on a scale from 0 to 14. Water that contains equal numbers of $H^+$ and $OH^-$ ions is considered neutral (pH 7). If a solution has more $H^+$ than $OH^-$ ions, it is considered acidic and has a pH less than 7. If the solution contains more $OH^-$ ions than $H^+$ ions, it is considered basic and has a pH greater than 7.

The pH scale is logarithmic. This means that as you go up and down the scale, the values change by a factor of 10. A pH change of one point indicates the strength of the acid or base has increased or decreased tenfold; a two-point pH change indicates a 100-fold change in acidity or basicity; a three point change indicates a 1000-fold change in acidity or basicity. For example, when the pH changes from 7 to 8, there are now 10 times more $OH^-$ ions present than $H^+$ ions. If it changes from 10 to 8, there are 100 times more $H^+$ ions than $OH^-$ ions. Small changes in pH result in large changes in water chemistry.

### HUMAN-CAUSED CHANGES IN PH

In the United States, the pH of rivers is usually between 6.5 and 8.5. Rain water is more acidic and normally has a pH between 5.0 and 5.6. Increased amounts of nitrogen oxides (NO3) and sulfur dioxide (SO2), primarily from automobile and coal-fired power plant emissions, are converted to nitric acid and sulfuric acid in the atmosphere, resulting in acid rain or snow. The geology of an area determines the pH of the local water. If limestone is present, like much of Missouri, the alkaline limestone neutralizes the effect acid rain might have on lakes and streams.

## CHANGES IN AQUATIC LIFE

Most organisms have adapted to life in water of a specific pH and may die if the pH changes. At extremely low or high pH values (4.5 or 11.0 respectively) the water becomes lethal to most organisms. pH is also important because of how it affects other pollutants in the water. Very acidic waters can cause heavy metals to be released into the water column. The metals can then be taken up and accumulated in the food chain. Metals in the water, such as copper and aluminum, can accumulate on the gills of fish or cause deformities in adolescent fish, reducing their chances of survival. In basic water, ammonia compounds convert to a toxic form; the more basic the water, the more toxic the ammonia that is present.

# 2

# Soil Classification

## THE ORIGIN OF SOIL

Soil is considered as the "skin of the earth" with interfaces between the lithosphere, hydrosphere, atmosphere, and biosphere. Soil is the mixture of minerals, organic matter, gases, liquids and a myriad of organisms that can support plant life. The constituents of the soil may be divided into the classes of organic matter and inorganic matter. Organic matters are derived from the decay of plants which have already grown upon the soil, and which, in various stages of decomposition, form the numerous classes of substances grouped together under the name of humus. The organic substances may therefore be considered as in a manner secondary constituents of the soil, which have been accumulated in it as the consequence of the growth and decay of successive generations of plants, while the primeval soil consisted of inorganic substances only. The inorganic constituents of the soil are obtained as the result of a succession of chemical changes going on in the rocks which protrude through the surface of the earth. We have only to examine one of these rocks to observe that it is constantly undergoing a series of important changes. Under the influence of air and moisture, aided by the powerful agency of frost, it is seen to become soft, and gradually to disintegrate, until it is finally converted into an uniform powder, in which the structure of the original rock is with difficulty, if at all distinguishable.

The rapidity with which these changes take place is very variable; in the harder rocks, such as granite and mica slate it is so slow as to be scarcely perceptible, while in others, such as the shales of the coal formation, a very few years' exposure is sufficient for the purpose. These actions, operating through a long series of years, are the source of the inorganic constituents of all soils.

Geology points to a period at which the earth's surface must have been altogether devoid of soil, and have consisted entirely of hard crystalline rocks, such as granite and trap, by the disintegration of which, slowly proceeding from the creation down to the present time, all the soils which now cover the surface have been formed.

But they have been produced by a succession of very complicated processes; for these disintegrated rocks being washed away in the form of fine mud, or at least of minute particles, and being deposited at the bottom of the primeval seas, have there hardened into what are called sedimentary rocks, which being raised above the surface by volcanic action or other great geological forces, have been again disintegrated to yield different soils. Thus, then, all soils are directly or indirectly derived from the crystalline rocks, those overlying them being formed immediately by their decomposition, while those found above the sedimentary rocks may be traced back through them to the crystalline rocks from which they were originally formed.

Such being the case, the composition of different soils must manifestly depend on that of the crystalline rocks from which they have been derived. Their number is by no means large, and they all consist of mixtures in variable proportions of quartz, felspar, mica, hornblende, augite, and zeolites. With the exception of quartz and augite, these names are, however, representatives of different classes of minerals. There are, for instance, several different minerals commonly classified under the name of felspar, which have been distinguished by mineralogists by the names of orthoclase, albite, oligoclase, and labradorite; and there are at least two sorts of mica, two of hornblende, and many varieties of zeolites.

Quartz consists of pure silica, and when in large masses is one of the most indestructible rocks. It occurs, however, intermixed with other minerals in small crystals, or irregular fragments, and forms the entire mass of pure sand. The four kinds of felspar which have been already named are compounds of silica with alumina, and another base which is either potash, soda, or lime.

It is obvious that soils produced by the disintegration of these minerals must differ materially in quality. Those yielded by orthoclase must generally abound in potash, while albite and labradorite, containing little or none of that element must produce soils in which it is deficient. The quality of the soil they yield is not however entirely dependent on the nature of the particular felspar which yields it, but is also intimately connected with the extent to which the decomposition has advanced.

It is observed that different felspars undergo decomposition with different degrees of rapidity but after a certain time they all begin to lose their peculiar lustre, acquire a dull and earthy appearance, and at length fall into a more or less white and soft powder.

During this change water is absorbed, and, by the decomposing action of the air, the alkaline silicate is gradually rendered soluble, and at length entirely washed away, leaving a substance which, when mixed with water, becomes plastic, and has all the characters of common clay.

In this instance the decomposition of the felspar had reached its limit, a mere trace of potash being left, but if taken at different stages of the process, variable proportions of that alkali are met with. This decomposition of felspar is the source of the great deposits of clay which are so abundantly distributed over the globe, and it takes place with nearly equal rapidity with potash and soda felspar.

It is rarely complete, and the soils produced from it frequently contain a considerable proportion of the undecomposed mineral, which continues for a long period to yield a supply of alkalies to the plants which grow on them. Mica is a very widely distributed mineral, and two varieties of it are distinguished by mineralogists, one of which is characterised by the large quantity of magnesia it contains.

Mica undergoes decomposition with extreme slowness, as is at once illustrated by the fact that its shining scales may frequently be met with entirely unchanged in the soil. Its persistence is dependent on the small quantity of alkaline constituents which it contains; and for this reason it is observed that the magnesian micas undergo decomposition less rapidly than those containing the larger quantity of potash.

Eventually, however, both varieties become converted into clay, their magnesia and potash passing gradually into soluble forms. Hornblende and augite are two widely distributed minerals, which are so similar in composition and properties that they may be considered together.

In these minerals alkalies are entirely absent, and their decomposition is due to the presence of protoxide of iron, which readily absorbs oxygen from the air, when the magnesia is separated and a ferruginous clay left.

The minerals just referred to, constitute the great bulk of the mountain masses, but they are associated with many others which take part in the formation of the soil. Of these the most important are the zeolites which do not occur in large masses but are disseminated through the other rocks in small quantity.

They are chiefly characterised by containing their silica in a soluble state, and hence, may yield that substance to the plants in a condition particularly favourable for absorption. It is obvious from what has been stated that all these minerals are capable, by their decomposition, of yielding soft porous masses having the physical properties of soils, but most of them would be devoid of many essential ingredients, while not one of them would yield either phosphoric acid, sulphuric acid, or chlorine.

It has, however, been recently ascertained that certain of these minerals, or at least the rocks formed from them, contain minute, but distinctly appreciable traces of phosphoric acid, although in too small quantity to be detected by ordinary analysis; and small quantities of chlorine and sulphuric acid may also in most instances be found. Still it will be observed that most of these minerals would yield a soil containing only two or three of those substances, which, as we have already learned, are essential to the plant.

Thus, potash felspar, while it would give abundance of potash, would be but an inefficient source of lime and magnesia; and labradorite, which contains abundance of lime, is altogether deficient in magnesia and potash. Nature has, however, provided against this difficulty, for she has so arranged it that these minerals rarely occur alone, the rocks which form our great mountain masses being composed of intimate mixtures of two or more of them, and that in such a manner that the deficiencies of the one compensate those of the other. We shall shortly mention the composition of these rocks. Granite is a mixture of quartz, felspar, and mica in variable proportions, and the quality of the soil it yields depends on whether the variety of felspar present be orthoclase or albite.

When the former is the constituent, granite yields soils of tolerable fertility, provided their climatic conditions be favourable; but it frequently occurs in high and exposed situations which are unfavourable to the growth of plants. Gneiss is a similar mixture, but characterised by the predominance of mica, and by its banded structure.

Owing to the small quantity of felspar which it contains, and the abundance of the difficulty decomposable mica, the soils formed by its disintegration are generally inferior. Mica slate is also a mixture of quartz, felspar, and mica, but consisting almost entirely of the latter ingredient, and consequently presenting an extreme infertility.

The position of the granite, gneiss, and mica slate soils in this country is such that very few of them are of much value; but in warm climates they not unfrequently produce abundant crops of grain. Syenite is a rock similar in composition to granite, but having the mica replaced by hornblende, which by its decomposition yields supplies of lime and magnesia more readily than they can be obtained from the less easily disintegrated mica.

For this reason soils produced from the syenitic rocks are frequently possessed of considerable fertility.

The series of rocks of which greenstone and trap are types, and which are very widely distributed, differ greatly in composition from those already mentioned. They are divisible into two great classes, which have received the names of diorite and dolerite, the former a mixture of albite and hornblende, the latter of augite and labradorite, sometimes with considerable quantities of a sort of oligoclase containing both soda and lime, and of different kinds of zeolitic minerals.

Generally speaking, the soils produced from diorite are superior to those from dolerite. The albite which the former contains undergoes a rapid decomposition, and yields abundance of soda along with some potash, which is seldom altogether wanting, while the hornblende supplies both lime and magnesia.

Dolerite, when composed entirely of augite and labradorite, produces rather inferior soils; but when it contains oligoclase and zeolites, and comes under the head of basalt, its disintegration is the source of soils remarkable for their fertility; for these latter substances undergoing rapid decomposition furnish the plants with abundant supplies of alkalies and lime, while the more slowly decomposing hornblende affords the necessary quantity of magnesia.

In addition to these, the basaltic rocks are found to contain appreciable quantities of phosphoric acid, so that they are in a condition to yield to the plant almost all its necessary constituents. The different rocks now mentioned, with a few others of less general distribution, constitute the whole of our great mountain masses; and while their general composition is such as has been stated, they frequently contain disseminated through them quantities of other minerals which, though in trifling quantity, nevertheless add their quota of valuable constituents to the soils.

Moreover, the exact composition of the minerals of which the great masses of rocks are composed is liable to some variety. Those which we have taken as illustrations have been selected as typical of the minerals; but it is not uncommon to find albite containing 2 or 3 per cent of potash, labradorite with a considerable proportion of soda, and zeolitic minerals containing several per cent of potash, the presence of which must of course considerably modify the properties of the soils produced from them.

They are also greatly affected by the mechanical influences to which the rocks are exposed; and being situated for the most part in elevated positions, they are no sooner disintegrated than they are washed down by the rains. A granite, for instance, as the result of disintegration, has its felspar reduced to an impalpable powder, while its quartz and mica remain, the former entirely, the latter in great part, in the crystalline grains which existed originally in the granite.

If such a disintegrated granite remains on the spot, it is easy to see what its composition must be; but if exposed to the action of running water, by which it is washed away from its original site, a process of separation takes place, the heavy grains of quartz are first deposited, then the lighter mica, and lastly the felspar.

Thus there may be produced from the same granite, soils of very different nature and composition, from a pure and barren sand to a rich clay formed entirely of felspathic debris. The sedimentary or stratified rocks are formed of particles carried down by water and deposited at the bottom of the primeval seas from which they have been upheaved in the course of geological changes.

The process of their formation may be watched at the present day at the mouths of all great rivers, where a delta composed of the suspended matters carried down by the waters is slowly formed. The nature of these rocks must therefore depend entirely on that of the country through which the river flows.

If its course runs through a country in which lime is abundant, calcareous rocks will be deposited, and if it passes through districts of different geological characters the deposit must necessarily consist of a mixture of the disintegrated particles of the different rocks the river has encountered. For this reason it is impossible to enter upon a detailed account of their composition.

It is to be observed, however, that the particles of which they are composed, though originally derived from the crystalline rocks, have generally undergone a complex series of changes, geology teaching that, after deposition, they may in their turn undergo disintegration and be carried away by water, to be again deposited.

Their composition must therefore vary not merely according to the nature of the rock from which they have been formed, but also according to the extent to which the decomposition has gone, and the successive changes to which they have been exposed. They may be reduced to the three great classes of clays, including the different kinds of clay slates, shales, etc., sandstone and limestone.

It must be added also, that many of them contain carbonaceous matters produced by the decomposition of early races of plants and animals, and that mixtures of two or more of the different classes are frequent. The purest clays are produced by the decomposition of felspar, but almost all the crystalline rocks may produce them by the removal of their alkalies, iron, lime, etc.

Where circumstances have been favourable, the whole of these substances are removed, and the clay which remains consists almost entirely of silica and alumina, and yields a soil which is almost barren, not merely on account of the deficiency of many of the necessary elements of plants, but because it is so stiff and impenetrable that the roots find their way into it with difficulty.

The sandstones are derived from the siliceous particles of granite and other rocks, and consist in many cases of nearly pure silica, in which case their disintegration produces a barren sand, but they more frequently contain an admixture of clay and micaceous scales, which sometimes form a by no means inconsiderable portion of them.

Such sandstones yield soils of better quality, but they are always light and poor. Where they occur interstratified with clays, still better soils are produced, the mutual admixture of the disintegrated rocks affording a substance of intermediate properties, in which the heaviness of the clay is tempered by the lightness of the sandstone. Limestone is one of the most widely distributed of the stratified rocks, and in different localities occurs of very different composition.

Limestones are divided into two classes, common and magnesian; the former a nearly pure carbonate of lime, the latter a mixture of that substance with carbonate of magnesia. But while these are the principal constituents, it is not uncommon to find small quantities of phosphate and sulphate of lime, which, however trifling their proportions, are not unimportant in an agricultural point of view.

These limestones are hard and possess to a greater or less extent a crystalline texture. They are replaced in later geological periods by others which are much softer, and often purer, of which the oolitic limestones, so called from their resemblance to the roe of a fish, and chalk are the most important. Other limestones are also known which contain an admixture of clay.

The soils produced by the disintegration of limestone and chalk are generally light and porous, but when mixed with clay, possess a very high degree of fertility, and this is particularly the case with chalk, which yields some of the most valuable of all soils. But it is true only of the common limestones, for experience has shown that those which contain magnesia in large quantity are often prejudicial to vegetation, and sometimes yield barren or inferior soils.

Such are the general characters of the three great classes of stratified rocks; any attempt to particularise the numerous varieties of each would lead us far beyond the limits of the present work. It is necessary, however, to remark, that in many instances one variety passes into the other, or, more correctly speaking, sedimentary rocks occur, which are mixtures of two or more of the three great classes.

In fact, the name given to each really expresses only the preponderating ingredient, and many sandstones contain much clay, shales and clay slates abound in lime, and limestones in sand or clay, so that it may sometimes be a matter of some difficulty to decide to which class they belong. Such mixtures usually produce better soils than either of their constituents separately, and accordingly, in those geological formations in which they occur, the soils are generally of excellent quality.

The same effect is produced where numerous thin beds of members of the different classes are interstratified, the disintegrated portions being gradually intermixed, and valuable soils formed. The fertility of the soils formed from the stratified rocks is also increased by the presence of organic remains which afford a supply of phosphoric acid, and which are sometimes so abundant as to form a by no means unimportant part of their mass.

They do not occur in the oldest sedimentary rocks, but as we ascend to the more recent geological epochs, they increase in abundance, until, in the greensands and other recent formations, whole beds of coprolites and other organic remains are met with. Great differences are observed in the quality of the soils yielded by different rocks.

In general, those formed by the disintegration of clay slates are cold, heavy, and very difficult and expensive to work; those of sandstone light and poor, and of limestone often poor and thin. These statements must, however, be considered as very general; for individual cases occur in which some of these substances may produce good soils, remarkable exceptions being offered by the lower chalk and some of the shales of the coal formation. Little is at present known regarding the peculiar nature of many of these rocks, or their composition; and the cause of the differences in the fertility of the soil produced from them is a subject worthy of minute investigation.

## PROPERTIES OF SOIL

No department of agricultural chemistry is surrounded with greater difficulties and uncertainties than that relating to the properties of the soil. When chemistry began to be applied to agriculture, it was not unnaturally supposed that the examination of the soil would enable us to ascertain with certainty the mode in which it might be most advantageously improved and cultivated, and when, as occasionally happened, analysis revealed the absence of one or more of the essential constituents of the plant in a barren soil, it indicated at once the cause and the cure of the defect.

But the expectations naturally formed from the facts then observed have been as yet very partially fulfilled; for, as our knowledge has advanced, it has become apparent that it is only in rare instances that it is possible satisfactorily to connect together the composition and the properties of a soil, and with each advancement in the accuracy and minuteness of our analysis the difficulties have been rather increased than diminished.

Although it is occasionally possible to predicate from its composition that a particular soil will be incapable of supporting vegetation, it not unfrequently happens that a fruitful and a barren soil are so similar that it is impossible to distinguish them from one another, and cases even occur in which the barren appears superior to the fertile soil. The cause of this apparently anomalous phenomenon lies in the fact that analysis, however minute, is unable to disclose all the conditions of fertility, and that it must be supplemented by an examination of its physical and other chemical properties, which are not indicated by ordinary experiments. Of late years very considerable progress has been made in the investigation of the properties of the soil, and many facts of great importance have been discovered, but we are still unable to assert that all the conditions of fertility are yet known, and the practical application of those recently discovered is still very imperfectly understood.

It must not be supposed that a careful analysis of a soil is without value, for very important practical deductions may often be drawn from it, and when this is not practicable it is not unfrequently due to its being imperfect or incomplete, for it is so complex that the cases in which all the necessary details have been eliminated are even now by no means numerous. In fact, the want of a large number of thorough analyses of soils of different kinds is a matter of some difficulty, and so soon as a satisfactory mode of investigation can be determined upon, a full examination of this subject would be of much importance.

## PHYSICAL PROPERTIES

The physical properties of a soil largely determine the manner in which it can be used. Properties, *e.g.,* the water holding capacity, permeability to water, aeration, plasticity and nutrient-supplying ability, are influenced by the size, proportion, arrangement and mineral composition of the soil particles.

The proportion of the four major components of the soils—inorganic or mineral particles, organic material, water and air—vary greatly from place to place and with depth. The amount of water air in a soil fluctuates from season to season, but the proportion of primary solid components of the soil, however remains unchanged.

## TEXTURAL CLASSES

The varying proportions of particles of different size groups in a soil constitute what is termed as a textural class. The principal textural classes are: Clay, sandy clay, silty clay, clay loam, sandy clay loam, silty clay loam, loam, sandy loam, silt loam, sand, loamy sand and silt. The physical properties and the chemical composition of the small and large particles differ greatly, the coarse fraction, gravel and sand which are mainly composed of rock fragments or primary minerals act as individual particles. They have low specific surface and are relatively non-reactive.

They do not hold large amounts of water or nutrients. Owing to large voids between them, they transmit air and water easily. The silt particles are intermediate between sands and clays. Mineralogically, the particles of silt are similiar to those of sand, as they are largely composed of primary minerals.

They are more reactive than sands because of the higher specific surface. The clay fraction controls of the important properties of the soils. They are chiefly composed of secondary minerals—crystalline alumino silicates. They have high specific surface and are most reactive. They have high capacity to retain water and nutrients. The textural classes differ not only in the particle size analysis, but also in their bearing on some of the important factors affecting plant growth,

Such as:

- The moveability and availability of water,
- Aeration,
- Workability, and
- The content of plant nutrients.

Sandy soils are very permeable and well drained but are less water retentive and hence, need more frequent irrigation for successful crop growth than fine textured soils. The clayey soils can hold more moisture, but they have high wilting percentage.

The rate of water intake of these soils is low. They are subject to water-logging resulting in poor aeration and workability. The moderately fine-textured soils, *e.g.,* loams, caly loams, or silt loams are by far the excellent soils for plant growth, since, they have the advantages of both sands and clays.

## DENSITY

Soils having larger particles are usually heavier in weight per unit volume than those having smaller particles. True density of a soil is based on the individual densities of soil constituents and according to their proportionate contribution.

The bulk density or apparent density is the weight per unit volume of dry soil as a whole, *i.e.*, particle and pore space and hence, it is lower than the true density. The relationship between the true density (T) and the apparent density (A) and the pore space (P) is as follows:

$$P \text{ per cent} = (T - A) \times 100/T.$$

In most mineral soils the true density varies within narrow limits of about 2.5 to 2.7 and the apparent density between 1.4 and 1.8.

## PORE SPACE

The pore space of soil is the portion occupied by air and water and it is determined largely by structural conditions. Sands have low pore space of about 30 per cent, whereas clays may have as much as 50–60 per cent.

Although clays possess greater total porosity than the sands the pore spaces in the latter being individually larger are more conducive to good drainage and aeration.

## PLASTICITY AND COHESION

Plasticity is the property that enables a moist soil to change shape on the application of force and retain this shape even when the force is withdrawn. On this basis, sandy soils may be considered to be non-plastic and clayey soils to be plastic.

Cohesion is the tendency of the particles to stick to ine another. Plastic soils are cohesive. Plasticity and cohesion reflect the soil consistency and workability of the soils.

## SOIL TEMPERATURE AND HEAT

Soil temperature is one of the important factors that control the microbiological activity and all the processes involved in the growth of plants. Heat is necessary for seed germination, root growth and other biological activities. The temperature needed for germination and root growth varies with crops and varieties.

Crops, *e.g.*, wheat, barley, and peas grown in India during winter germinate at relatively low temperatures as compared with maize, and those at which groundnut or cotton germinate. Microbiological activities are retarded by low soil temperature. As a result, the nitrification processes in the soil are slowed down and plant nutrition and growth adversely affected. Soil temperature is to be considered so important that in the soil taxonomy there is a provision to use it as a differentiating criterion at the family level of categorisation.

## SOIL AIR

It is well known that restricted soil aeration adversely affect root development, processes of respiration and other essintial biological processes. It is therefore, important to know the content of the soil air and its composition. It has already been mentioned that, depending on its texture, the soil may have pore space of 30–60 per cent. The pore space not fill by water is occupied by air.

Under moist field conditions, the non-capillary pore space generally constitutes the air space, the capillary pore space being occupied by water. Ordinarily, the occupation of nearly one-third of the pore space in the soil by air and two-thirds of it by water constitutes the most favourable condition for the plant growth.

The composition of soil air reflects a dynamic balance between two competing processes. The consumption of oxygen and the liberation of carbon dioxide by plant roots and the soil organisms tend to increase the difference in the composition of the soil air and the atmosphere above the soil surface gaseous diffusion tents to reduce the difference in composition.

## SOIL STRUCTURE

Soil stricture refers to the arrangement of soil particles, both primary and secondary. Soil structure is one of the most important properties of soil mass, since, it influences aeration, permeability, water capacity, etc.

*In the field the structure is described in terms of:*

- Type-referring to shape and arrangement
- Class-referring to size
- Grade-referrint to the degree of aggregation.

*There are 4 types of primary structures:*

1. *Platy:* With particles arranged around a plane, generally horizontal.
2. *Prism-like:* With particles arranged around a vertical axis and bounded by relatively flat vertical surfaces which are the casts or the molds formed by the faces of the surrounding peds. This type includes prismatic and columnar types.
3. *Block-like:* With particles arranged around a point and bounded by flat or rounded surfaces which are the casts or the molds formed by the faces of the surrounding peds. This type includes angular and subangular types.
4. *Spheroidal:* Particles arranged around a point and bounded by curved or very irregular surfaces that are not accomodated in the adjoining aggregates. This type includes granular and crumb types.

## CLASSES

Five size classes are recognised in each of the primary types. They are fine, very fine, medium, coarse and very coarse. The actual size for classes in each type varies.

## GRADE

The grade of structure representing the degree of aggregation is determined in the field mainly by noting the durability of the aggregates and the proportions between aggregated and unaggregated material that results when the aggregates are disturbed or gently crushed. Grades are termed structureless (massive if coherent, and single-grained if non-coherent), weak, moderate, strong and very strong, depending on the stability of aggregates when disturbed.

## FACTORS AFFECTING STRUCTURE

Structure is primarily influenced by texture. The beneficial effects of organic matter on the structure and working of the soils are well known. Soil structure is also influenced by:

*Soil Management:* The cutting action of ploughs or other tillage implements breaks up the soil mass and may have favourable or adverse effects on the structure, depending on whether the soil is worked under optimum moisture conditions or not.

A good soil management, with proper system of crop rotation has the effect of maintaining the soil in a good state of aggregation.

*Absorbed Cations:* Sodium and potassium ions on the clay complex have a tendency to disperse the soil; calcium has favourable effects on the aggregation. Similiarly the presence of soluble salts favours floculation.

*Micro-Organisms:* The filamentous growth of soil fungi and the microbial decomposition products of organic matter have a binding effect on soil particles thus favouring aggregation. Similiarly the excrements of earthworms and the burrowing activities of insects and other creatures cause appreciable changes in the soil structure.

*Variations of Soil Moisture:* Variations of soil moisture due to drying and wetting influence the structure. The drying of soil forms cracks and big clods. Poorly drained soils with excess of moisture usually have an unfavourable structure. The structure of cultivated soils have a very important bearing on their drainage, ease of tillage, root penetration and resistance to erosion and ultimately their general productivity.

From this point of view the crumb and granular structure (spheroidal) is considered very favourable to plant growth. The structure and physical properties of soil can be improved by adopting a suitable system of soil management, including legumes in the rotation system, manuring, and regularly supplying the soil with organic manure.

## MINERAL COMPOSITION OF THE SOIL SEPARATES

Owing to a varying rate if wearing down of different minerals, the more easily decomposing feldspars predomonate in the relatively fine separates, whereas the resistant minerals, *e.g.,* are more abundant in coarser separates. The separates therefore show differences in chemical composition also. As indicated earlier, the clay which has a high specific surface is more reactive than silt or sand.

### THE CLAY FRACTION

In the process of decomposition relatively stable new elements are formed from the products of weathering which constitute largely the clay fraction.

The clay particles formed in this manner are crystalline and as shown by X-ray diffraction and ptrographic techniques are composed of sheets of hydrated alumina and silica linked by oxygen atoms.

Because of their being very small, they are highly reactive and form the seat of ion exchange in the soil and this controls and regulates adsorption, retention, and release of many plant nutrients, such as potassium, calcium, magnesium and phosphorous.

The precise nature of clay properties depends on the type of minerals that predominantly compose the clay. These fall into three main groups:

*Kaolinite:* A unit kaolinite crystal lattice consists of one sheet of silica and one sheet of alumina and is hence, called 1:1-layer silicate. The two sheets are held together by mutually shared oxygen atoms.

These units in turn are tenaciously held together by oxygen-hydroxyl linkages and cosequently no expansion occurs when the mineral is wetted. It has a low specific surface and a low caton exchange capacity. Similiarly plasticity, cohesion, shrinkage and the swelling properties of kaolinite are low.

*Montmorillonite:* A unit montmorillonite crystal lattice consists of 2 sheests of silica and 1 sheet of alumina (2:1-layer silicate) held together by mutually shared oxygen atoms. Such units however are held together by weak oxygen-oxygen linkages.

There is some isomorphic substitution of iron or magnesium in the alumina sheet; montmorillonite crystals can expand and owing to this expansion, cations and water molecules are able to move in between the crystal units. Montmorillonite has a high specific surface and caton-exchange capacity. Similiarly, plasticity, cohesion, shrinkage and swelling properties are high.

*Iillite:* Illite has the same general structure organisation as montmorillonite except in respect of the linkages between the crystal units. About 15 per cent of silica in the silica sheet is replaced by aluminium and potassium atoms and they supply the additional connecting linkages between the crystal units, due to which illite shows a lower expansion capacity, it has properties intermediate between kaolinite and montmo-rillonite.

## SOIL COLLOIDS

The most active portions of the soil are those which are in the colloidal state. The colloidal state implies a two-phase system in which the material called the dispersed phase (fine clay and humus) is dispersed in the dispersed medium(water). In soils, the mineral and organic colloids exist in intimate and heterogenous admixture. The mineral colloid is present almost exclusively as the clay of various kinds, whereas the organic colloid is present as humus. It is generally supposed that clay particles less than one micron in diameter possess colloidal properties and these properties increase with a decrease in the size of the particles.

*The most distinctive colloidal properties are:*

- The large specific surface or interface, and
- The capacity to hold solids, gases, salts and ions.

Depending on their nature, the soil colloids determine many of the physical and chemical properties. The colloidal material may occur as a thin gelatinous film around coarser particles, or it may occupy a considerable part of the space between the larger particles, thus serving as a binding material.

Soil colloids especially the mineral colloids, exhibit cohesive and adhesive properties. Soils with a high amount of colloidal clay compete with plant roots for water and mineral nutrients, especially at lower levels of their availability. Generally the soil colloids have a high exchange capacity, which increases with silica sesquioxide ratio.

# CHEMICAL PROPERTIES

## MINERAL MATTER AND ITS COMPOSITION

The principal minerals occurring in the earth's crust are:

| Mineral | Approximate Percentage |
|---|---|
| Feldspars | 48 |
| Quartz | 36 |
| Micas | 10 |
| Limestone and dolomite | 2 |
| Horneblende and augite | 1 |
| Clays | 1 |
| Other minerals | 1 |

*Feldspars:* Feldspars are anhydrous alumino silicates of potassium, sodium and calcium:

- *Quartz*: Quartz or silica is silicon dioxide found in crystalline rocks. It also occurs as free silica or sand.

- *Mica*: Mica is layer of alumino silicate of potassium; iron, magnesium and sodium may be present in varying proportions.
- *Limestone*: It is largely calcium carbonate; with magnesium carbonate it occurs as dolomite.
- *Hornblende And Augite*: They are the ferro magnesium minerals consisting of silicates of calcium, magnesium, iron and sodium in varying proportions.
- *Olivine and Serpentine*: Olivine is an olive-ferro-magnesium silicate. Serpentine is a hydrated silicate of magnesium.
- *Clays*: Clays are secondary minerals. They are hydrated alumino silicates.
- *Other Minerals*: Other minerals occurring in the soil include tourmaline- a boro alumino silicate with alkali metals and iron or magnesium; rutile-titanium oxide, zircon-ziconium silicate, glauconite-hydrated silicate of iron and potassium, apatite-calcium phosphate, sulphur-bearing minerals, etc. Allophane and the hydrous oxides of aluminium, iron and titanium also commonly occur in soils and more so in highly weathered and leached soils.

## INORGANIC COMPONENTS

From what has been stated in the preceding section it is evident that compounds of silicon, aluminium, calcium, magnesium, iron, potassium and sodium form the principal chemical constituents of the mineral matter in the soils. Besides, the soil contains small amounts of a large number of other mineral elements, *e.g.,* phosphorus, boron, manganese, copper, sulphur, zinc and cobalt.

Thus the soil supplies all the following essential mineral elements required by the plants:

- *Macros*: Phosphorus, potassium, calcium, magne-sium and sulphur.
- *Micros*: iron manganese, zinc, copper, molybdenum, boron and chlorine. Plants obtain carbon, hydrogen and oxygen from carbon dioxide and water. Soil is also the source of nitrogen for plants, but the ultimate source of nitrogen is traceable to atmosphere from where nitrogen fixation takes place physico-chemically and biologically.

The total amount of elements contained in the soils depends partly on the nature of the parent material from which they are formed and partly on their age and extent to which soluble products have been leached down. The chemical composition of different horizons of a soil also show a good deal of variation. Generally, ‘A’ horizon is richer in soluble components than the ‘B’ horizon. Usually some of the elements that are commonly leached out are also ones that are required by the plants. The percentage inorganic composition of eight representative Indian soils and OS the successive horizons of the soils.

## CHEMICAL COMPOSITION OF THE SOIL

Reference has been already made to the division of the constituents of the soil into the two great classes of organic and inorganic. And when treating of the sources of the organic constituents of plants, we entered with some degree of minuteness into the composition and relations of the different members of the former class, and expressed the opinion that they did not admit of being directly absorbed by the plant.

But though the parts then stated lead to the inference that, as a direct source of these substances, humus is unimportant, it has other functions to perform which render it an essential constituent of all fertile soils. These functions are dependent partly on the power which it has of absorbing and entering into chemical composition with ammonia, and with certain of the soluble inorganic substances, and partly on the effect which the carbonic acid produced by its decomposition exerts on the mineral matters of the soil.

In the former way, its effects are strikingly seen in the manner in which ammonia is absorbed by peat; for it suffices merely to pour upon some dried peat a small quantity of a dilute solution of ammonia to find its smell immediately disappear. This peculiar absorptive power extends also to the fixed alkalies, potash and soda, as well as to lime and magnesia, and has an important effect in preventing these substances being washed out of the soil—a property which, as we shall afterwards see, is possessed also by the clay contained in greater or less quantity in most soils.

On the other hand, the air and moisture which penetrate the soil cause its decomposition, and the carbonic acid so produced attacks the undecomposed minerals existing in it, and liberate the valuable substances they contain.

In considering the composition of a soil, it is important to bear in mind that it is a substance of great complexity, not merely because it contains a large number of chemical elements, but also because it is made up of a mixture of several minerals in a more or less decomposed state. The most cursory examination shows that it almost invariably contains sand and scales of mica, and other substances can often be detected in it.

Now it has been already observed that the minerals of which soils are composed, differ to a remarkable extent in the facility with which they undergo decomposition, and the bearing of this fact on its fertility is a matter of the highest importance, for it has been found that the mere presence of an abundant supply of all the essential constituents of plants is not always sufficient to constitute a fertile soil. Two soils, for instance, may be found on analysis to have exactly the same composition, although in practice one proves barren and the other fertile.

The cause of this difference lies in the particular state of combination in which the elements are contained in them, and unless this be such that the plant is capable of absorbing them, it is immaterial in what quantity they are present, for they are thus locked up from use, and condemn the soil to hopeless infertility.

It is admitted that unless the substances be present in a state in which they can be dissolved, the plant is incapable of absorbing them; but it is a matter of doubt whether it is necessary that they be actually dissolved in the water which permeates the soil, or whether the plant is capable of exercising a directly solvent action.

The latter view is the most probable, but at the same time it cannot be doubted, that if they are presented to the plant in solution, they will be absorbed in that state in preference to any other. Hence, it has been considered important in the analysis of a soil, not to rest content with the determination of the quantity of each element it contains, but to obtain some indication of the state of combination in which it exists, so as to have some idea of the ease or difficulty with which they may be absorbed.

*For this purpose it is necessary to determine:*

- *1st,* The substances soluble in water;
- *2nd,* Those insoluble in water, but soluble in acids;
- *3rd,* Those insoluble both in water and acids; and if to these the organic constituents be added, there are four separate heads under which the components of a soil ought to be classified.

This classification is accordingly adopted in the most careful and minute analyses; but the difficulty and labour attending them has hitherto precluded the possibility of making them except in a few instances; and, generally speaking, chemists have been contented with treating the soil with an acid, and determining in the solution all that is dissolved.

Such analyses are often useful for practical purposes, as for example, when they show the absence of lime, or any other individual substance, by the addition of which we may rectify the deficiency of the soil; but they are of comparatively little scientific value, and throw but little light on the true constitution of the soil, and the sources of its fertility.

Nor is it likely that much satisfactory information will be obtained until the number of minute analyses is so far extended as to establish the fundamental principles on which the various properties of the soil depends.

The separation of the constituents of a soil into the four great groups already mentioned, is effected in the following manner:—A given quantity of the soil is boiled with three or four successive quantities of water, which dissolves out all the soluble matters. These generally amount to about one-half per cent of the whole soil, and consist of nearly equal proportions of organic and inorganic substances.

In very light and sandy soils, it occasionally happens that not more than one or two-tenths per cent dissolve in water, and in peaty soils, on the other hand, the proportion is sometimes considerably increased, principally owing to the abundance of soluble organic matters. When the residue of this operation is treated with dilute hydrochloric acid, the matters soluble in acids are obtained in the fluid.

The proportion of these substances is liable to very great variations, and in some soils of excellent quality, and well adapted to the growth of wheat, it does not exceed 3 per cent; while in calcareous soils, such as those of the chalk formation, it may reach as much as 50 or 60 per cent. In general, however, it amounts to about 10 per cent.

The organic constituents are also very variable in amount; ordinary soils of good quality containing from 2 to 10 per cent, while in peat soils they not unfrequently reach 30 or even 50 per cent. But these cannot be considered *fertile* soils. The insoluble constituents are likewise subject to great variations, but, in the ordinary clay and sandy soils of this country, they generally form from 70 to 85 per cent of the whole.

The distribution of the constituents under these different heads will be best illustrated by a few analyses of soils of good quality, and for this purpose we shall select two, noted for the excellent crops of wheat they produce, and for their general fertility. The analyses were made from the upper 10 inches, and a quantity of the 10 inches immediately subjacent was analysed as subsoil.

The first is the ordinary wheat soil of the county of Mid-Lothian, the other the alluvial soil of the Carse of Gowrie in Perthshire, so celebrated for the abundance and luxuriance of the crops it produces.

In examining these analyses, it is particularly worthy of notice that by far the larger proportion of the substances soluble in water consists of organic matter, lime, and sulphuric acid, the two last being in combination as sulphate of lime, while some of those substances which are usually considered to be the most important mineral constituents of plants are present in very small quantity—potash, for instance, forming not more than 1-25,000th of the whole soil, and phosphoric acid being entirely absent.

On the other hand, this portion contains the whole of the chlorine which exists in the soil, and this might be anticipated from the ready solubility in water of the compounds of that substance. The portion soluble in acids consists of alumina and oxide of iron, both of which are comparatively unimportant to the plant, but very important, as we shall afterwards see, in relation to the physical properties of the soil.

The remainder of the substances soluble in acids, amounting to from 1 and 2 per cent, is composed of some of the most essential constituents of plants. Lime, magnesia, potash, and soda, appear again in larger quantity than in the soluble part, and along with them we have the phosphoric acid to the amount of from 0·2 to 0·4 per cent of the whole soil, and sulphuric acid in much smaller quantity.

The insoluble matters differ remarkably in the two soils, that from the Carse of Gowrie being characterised by a large quantity of potash and soda, indicating an important difference in the materials from which they have been formed. In the Perthshire soil it is obvious that the felspathic element has been abundant, and that its decomposition has been arrested at a time, when it still contained a large quantity of alkalies.

And this difference is of great practical importance, because those soils, which contain a large quantity of potash in their insoluble portion, have within them a source of permanent fertility, the alkali being gradually liberated by the decomposition which is constantly in progress, owing to the air and moisture permeating the soil.

As regards the special distribution of the inorganic matters, it is to be observed that some of them occur in each of the three heads under which they are arranged, while others are confined to one or two. Silica and the alkalies occur generally, though not invariably, in all three.

Chlorine is met with only in the part soluble in water, phosphoric acid only in that soluble in acids, while sulphuric acid occurs in both the last-named divisions. The greater part of the organic matters are insoluble both in water and acids. At least it is generally believed that any portion dissolved by strong acids, in the course of analysis, has been entirely decomposed, and is in a completely different state from that in which it existed actually in the soil.

As an example of a calcareous soil, forming a striking contrast to those given above, we select one from the island of Antigua, from which very large crops of sugar-cane are obtained. The soil is of great depth, and analyses of the subsoil at the depth of 18 inches and 5 feet are given. These last analyses are not so minute as that of the soil itself, the soluble matters not having been separately determined, but included in that soluble in acids.

In this soil there is a general resemblance in the composition of the portion soluble in water to those of the wheat soils. But the part soluble in acids is distinguished by the great abundance of carbonate of lime. The subsoil contains also a large quantity of protoxide of iron, a substance frequently found in subsoils containing much organic matter, and to which the air has imperfect access.

Under these circumstances peroxide of iron is reduced to protoxide; and when present abundantly in the soil in that form, iron has been found to exercise a very injurious influence on vegetation; and it has frequently happened that when subsoils containing it have been brought up to the surface, they have in the first instance caused a manifest deterioration of the soil, although after some time, when it had become peroxidised by the action of the air, it ceased to be injurious.

It is unnecessary to multiply analyses of fertile soils, those now given being sufficient to show their general composition. They are all characterised by the presence, in considerable quantity, of all the essential constituents of plants, in a state in which they may be readily absorbed. The absence of one or more of these substances immediately diminishes or altogether destroys the fertility of the soil; and the extent to which this occurs is illustrated by the following analysis of a soil from Pumpherston, Mid-Lothian, forming a small patch in the lower part of a field, and on which nothing would grow. Being naturally wet, it had been drained and sowed with oats, which died out about six weeks after sowing, and left a bare soil on which weeds did not show the slightest disposition to grow.

## SOIL SURVEY AND MAPPING

Modern soil surveys are basically aimed at fuenishing comprehensive information about soils and an inventory of soil resources from that area. They cosist in studying and recording important characteristics of the soils in the field and in the laboratory classifying them into well defined units and locating their extent and boundaries on a map. The broad general objectives of soil surveys are both of fundamental and applied nature.

Fundamentally, soil surveys help to expand our knowledge and understanding of soils as regards their genesis, development, classification and nomenclature. The applied part of the survey includes the interpretation of soil data for use in agriculture, forestry, engineering, urban development, etc. it gives information on how to evolve a rational land use plan in a given area.

Further it helps corelate the characteristics of soils of known behaviour and predict their adaptibility to various uses and also their behaviour and productivity under a defined set of management practices. It thus forms the very basis for planned land use.

*Types of Surveys:* Two main types of surveys are:

1. Reconnaissance survey, and
2. Detailed surveys, depending upon the method, the intensity of the survey, the scale of the base map, and the resulting details and precision of mapping.

*Reconnaissance Soil Survey:* In the case of a reconnaissance survey, the scale used for mapping is 1:63,360 (1" = 1 mile) survey of India toposheet, or 1:50,000 or a similiar scale air photographs, wherever they are available, mapping is done after an initial interpretation of the base maps and by constructing the initial legend.

Soil series or an association of soil series is normally the unit of mapping. The intensity of profile examination or the auger-bore observation depends on the heterogeneity of the land and soil features. In a highly heterogenous area the number of observations for mapping would be large and vice-versa.

Reconnaissance surveys and maps serve important purposes of identifying bench-mark soils, delineating problem and potential areas and providing information useful for a broad land use planning and agricultural development at a tehsil level and upwards, progressive abstraction, synthesis and the compilation of such tehsil maps lead to the generalised district, state and country level soil maps showing the delineation and classification soils at progressively higher categories.

*Detailed Soil Survey:* In this case, maps with a scale lerge enough to delineate the phases of soil types, slopes, erosion, etc., within a series are used. Cadestral maps on a scale of about 1:8000 or larger, or aerial photographs of 1:20,000 or larger are used as base maps. Depending upon the land and soil heterogeneity and the extent of details required the profile and auger-bore observations are made at closer intervals and the mapping units at the phase level are delineated.

The detailed soil map furnishes information needed for understanding the soil and land problems and for working out measures needed for the conservation of soil, for water-use management, drainage, etc. and the maintenance of soil fertility without deterioration.

## SOIL SURVEY INTERPRETATION

*Land Capability Classification:* The land-capability classification is an interpretative grouping of soils primarily meant for agricultural purposes and is mainly based on:

- The inherent soil characteristics
- The external land features
- Environmental factors.

The classification of soil units mapped into capability groupings enables one to understand:

- The hazards of the soil to various factors which cause soil damage, deterioration in fertility and
- Its potentiality for production.

The capability classification provides 3 major categories:

1. *The capability unit*: It is a grouping of soils that have almost the same influence on production and responses to the systems of management of common cultivated crops.
2. *The capability sub-class*: It is a grouping of capability units having similiar limitations and hazards. Four kinds of limitations recognised are:
   - Erosion,
   - Wetness,
   - Shallowness of root zone and
   - Climate.
3. *The Capability class*: It is the broadest category of classification. 8 classes are recognised and the risk of soil damage or limitations in

use progressively increases from class 1 to class 8. Soils in classes 1 to 4 are arable. They are capable of producing commonly cultivated crops of the region under good management. Soils in the classes 5 to 7 are suited to adapted native plants, pasture or forestry. Soils of class 8 are neither suited for agriculture nor for silviculture. The capability classes are briefly described below:

**ARABLE LAND**

- *Class I*: These soils have only a few limitations that restrict their use. They soils are nearly level, deep, well drained with good water holding capacity. They are productive and suitable for intensive cropping.
- *Class II*: These soils have some limitations that reduce the choice of crops and require moderate conservation practices to prevent deterioration when cultivated. The soils may be used for raising many of the cultivated crops.
- *Class III*: These soils have severe limitations that reduce the choice of crops and require special conservation practices when used for raising cultivated crops. The limitations are more than in class II and thus need careful management.
- *Class IV*: These soils have very severe limitations that reduce the choice of crops and need very careful management. The cultivation of crops may be restricted to once in 3-4 years. Conservation measures are more difficult to apply and maintain in these soils.

## CHEMICAL PROBLEMS OF SOILS

All landuse activities, particularly those which are poorly managed, involve destruction or disturbance, to a greater or lesser extent, of natural and semi-natural ecosystems. Almost invariably, however, it is these ecosystems, in equilibrium with their environment, which offer most effective protection to the soil which supports them. A major consequence of ecosystem destruction and disturbance is that of soil degradation.

This has been defined as the decline in soil quality caused through its misuse by human activity. More specifically, it refers to the decline in soil productivity through adverse changes in nutrient status, organic matter, structural stability and concentrations of electrolytes and toxic chemicals. Soil degradation incorporates a number of environmental problems, some of which are interrelated, including erosion, compaction, water excess and deficit, acidification, salinisation and sodification, and toxic accumulation of agricultural chemicals and urban/industrial pollutants. In many instances, these have led to a serious decline in soil quality and productivity, and it is only in recent decades that the finite nature of soil as a resource has become widely recognised.

Soil degradation is not a new phenomenon. Archaeological evidence suggests that it has been on-going since, the beginning of settled agriculture several thousand years ago. The decline of many ancient civilizations, including the Mesopotamians of the Tigris and Euphrates valleys in Iraq, the Harappans of the Indus valley in Pakistan and the Mayans of Central America, was due in part to soil degradation. More recently, an event of major significance was the dustbowl which occurred in the Great Plains of the American midwest during the 1930s.

At this time, intensive agricultural practices, employed in the eastern states, were transferred to the drier midwest where the soils are lighter textured and more susceptible to erosion. A number of years of drought, combined with crop failure and destruction of the protective organic-rich topsoil, resulted in severe wind erosion.

The effects of soil degradation are not restricted to the soil alone, but have a number of off-site implications. Soil erosion, for example, is often associated with increased incidence of flooding, siltation of rivers, lakes and reservoirs, and deposition of material in low-lying areas. These problems may be compounded in areas where infiltration capacity is reduced due to compaction, hardsetting or induration of soils.

Salinisation and sodification of soils are often associated with poor quality irrigation water, while soil acidification is commonly linked with acidification and aluminium contamination of surface waters. Leaching of fertilizers and pesticides from agricultural soils may also lead to contamination of surface and shallow ground waters. In addition, contamination of soils by urban and industrial pollutants, such as heavy metals and radionuclides, may lead to toxic accumulation in arable produce and in herbage for grazing animals, thus having important implications for human health.

The extent of soil degradation is influenced by a number of factors, many of which are interrelated, namely soil characteristics, relief, climate, land use, and socio-economic and political controls. In many studies of soil degradation and its wider environmental implications, the socio-economic and political controls are often overlooked, or at least not examined in any detail, perhaps because of the difficulties associated with the collection of reliable and comparable data.

Increasingly, however, these controls on land use systems are being viewed as central to the issue of soil degradation, particularly in the developing world. Management of soil degradation, whether at a global, regional or local scale, is clearly a complex issue and represents one of our most challenging environmental problems.

Emphasis should be placed on sustainable rather than exploitative landuse practices; this theme was highlighted by the World Soil Charter which called for a commitment by governments, agencies and land users to 'manage the land for long term advantage rather than short term expediency'. The problem requires a holistic, multidisciplinary approach involving the collaborative and co-ordinated efforts of ecologists, agronomists, soil scientists, hydrologists, engineers, sociologists and economists.

Moreover, the involvement of government and non-government organisations, aid agencies and the farmers themselves is essential to the success of research and development in this area. Such involvement should facilitate the implementation of education, training and incentive programmes.

Imposition from above of high-technology, high-cost solutions by technical experts from developed countries is certainly not the answer in the developing world. Inevitably such solutions are not economically viable and low-technology, low-cost options, such as low external input agriculture, agroforestry and social forestry, are often the only answer.

Hence, the approach to soil conservation has shifted in recent years from a rather technocentric standpoint to a more ecocentric position. Central to this approach are the concepts of land husbandry and sustainable development, which place emphasis on the land users themselves rather than on the technical experts and advisors.

The most pressing soil degradation problems, and in each case the causal factors, on- and off-site effects, and management strategies will be considered.

*Compaction:* Soil compaction involves the compression of a mass of soil into a smaller volume and is usually expressed in terms of dry bulk density, porosity and resistance to penetration. The bulk density of soils which have been compacted by agricultural machinery, for example, may exceed 1.5 g/cm$^3$, while that of comparable uncompacted soils is usually between 1.0 and 1.5 g/cm$^3$.

The ease with which soils are compacted depends on a number of characteristics, particularly texture, occurring most readily in soils which contain appreciable quantities of clay. The compacting force, which usually acts in a vertical direction, causes alignment of the clay platelets in a direction which is more or less parallel to the ground surface. Alignment of the clay particles in this way often leads to the formation of compaction or cultivation pans which may be a few centimetres in thickness.

Such pans tend to form at a depth of about 20–30 cm, are often characterised by a well developed platy structure, and are commonly associated with impeded drainage and restricted development of plant root networks. In silty soils, raindrop impact may lead to surface crusting which is another form of soil compaction.

Surface crusts may be several millimetres in thickness, and are commonly associated with reduced infiltration, increased surface run-off and accelerated soil erosion, and restricted germination and emergence of crops. In addition to textural controls, susceptibility to compaction also increases with increasing organic matter and water contents. Closely related to soil compaction is the process of hardsetting which involves an increase in bulk density but without the application of an external load. Hardsetting soils are characterised by their low structural stability. During and after wetting, slaking and collapse of aggregates lead to uniaxial shrinkage and dispersal of clay and silt and, on drying, the soils harden without restructuring. Acceptance of hardsetting soils as a distinctive group has been hindered by the difficulties experienced in distinguishing hardsetting characteristics from other forms of soil behaviour.

One of the main causes of soil compaction is the excessive use of heavy agricultural machinery, and compaction is particularly severe along wheelings left by these vehicles. For example, bulk density values of 2.2 g/cm$^3$ and 1.3 g/cm$^3$ have been recorded in soils of wheeled and inter-wheel areas respectively. The intensification of arable cultivation in recent decades has been facilitated to a large extent by increased mechanisation.

Most procedures in the cropping cycle, from tillage and seedbed preparation, through drilling, weeding and agrochemical applications to harvesting, are now largely mechanised, particularly in the developed world. The timing of cultivation in relation to precipitation and soil water levels is known to be critical with respect to soil compaction. If soils are cultivated when they are near or above field capacity, or their plastic limit, then severe structural damage and compaction are likely to occur.

Soil compaction is not restricted to arable land but is also common on land used for grazing and forestry. Grazing animals are well known for their ability to cause compaction, or poaching, especially during wet conditions.

It is particularly common to find poorly drained or puddled areas in the vicinity of food troughs or gateways where animals tend to congregate. In forestry plantations in upland Britain, heavy agricultural machinery is often used during the preparation and drainage of land prior to planting, and during harvesting operations; such mechanisation is often associated with the disturbance and compaction of soils. Similarly, soil compaction has been reported in association with the clearance of tropical rainforest, particularly where mechanical methods are used in preference to the traditional slash and burn approach.

Soil restoration following mineral extraction, quarrying and opencast coal mining has also been implicated in the soil compaction problem. Compacted horizons result largely from the passage of heavy machinery during soil stripping and replacement, and from the shear forces produced during the lifting process.

Compaction has adverse effects on a number of soil characteristics. Increased bulk density and the consequent decrease in porosity are associated with both increased water logging and poor aeration; these changes in turn have a detrimental effect on the thermal characteristics of soils.

Under these conditions, plant growth may be restricted, particularly during the early stages of germination and emergence, and during the main phase of root network development. Similarly, the increased incidence of root rot in soils with compacted cultivation pans has been widely documented. Other characteristics that may be adversely affected by compaction include the size and diversity of soil organism populations, and the incidence of certain crop pests and diseases may also increase.

The financial impact of soil compaction is difficult to assess due to the influence of a number of interrelated factors. In addition to yield reductions, damaged soil structure may have to be restored, surface run-off and soil erosion may increase, fertilizer usage becomes less efficient due to increased leaching losses, water loss by evaporation may increase, together with the operational costs of irrigation, and in soils which are susceptible to water logging, expensive drainage systems may have to be installed. It is estimated that the cost of fuel required for tillage of compacted soils may be up to 35 per cent greater than that for uncompacted soils.

Successful management and remediation of soil compaction centres on two main approaches—improvement and maintenance of soil structure, and appropriate tillage practices. Good soil structure provides the basis for increased mechanical strength and stability, improved water retention and aeration, and protection of plant nutrients from leaching. Soil structure can be improved and maintained through the promotion of sensible crop rotation practices, particularly those which include significant periods of ley pasture, mulching and stubble retention.

Unlike continuous cropping, such practices encourage the accumulation of soil organic matter and the development of strong plant root networks, both of which are essential ingredients in the development of good soil structure. Soil structure also benefits from the addition of certain nutrients, particularly the multivalent base cations which play a crucial role in the aggregation process.

In addition to soil structural improvements, good tillage practices are an essential component of effective management and remediation of soil compaction. Tillage of agricultural land is practised for a number of reasons—first, to create a fine tilth which facilitates effective germination and emergence of seedlings, second, to remove weeds which compete with crops for nutrients, water and light, third, to incorporate crop residues into the soil with the aim of improving soil structure and nutrient content, and fourth, to improve drainage and aeration in the root zone (rhizosphere).

Although tillage is therefore of benefit to the soil, when mechanised it can lead to soil compaction. The timing of tillage is particularly important in this respect, and it should be avoided in wet conditions, particularly if soil moisture content is likely to exceed the field capacity or plastic limit.

In an attempt to alleviate compaction resulting from the excessive use of heavy agricultural machinery, reduced or zero-tillage practices have been widely adopted. Such practices are inappropriate for poorly drained soils, however, where conventional tillage helps to improve drainage, aeration and the thermal characteristics of the topsoil.

They should also be avoided on excessively drained soils, where increased nitrate leaching and denitrification may lead to nitrogen deficiency, and on steep slopes where seed slots may be subjected to increased erosion.

Thus, it appears that the benefits of reduced and zero-tillage are felt most strongly on medium-textured soils which are well drained and which possess a well developed structure. In some circumstances, compaction and associated poor drainage cannot be improved through good structural management and appropriate tillage practices alone, and installation of artificial drainage may be necessary.

*Water Excess and Deficit:* Excess water may be present in soils for a number of reasons. Compaction, for example, in the form of surface crusting, subsurface cultivation pans and hardsetting, often leads to reduced infiltration and restricted permeability. Excess water is also common in heavily textured clay soils, particularly in high rainfall areas, and in areas which are flat, low lying and prone to flooding. If excess water is found within about 40–50 cm of the surface, then the adverse effects on soil aeration and temperature are likely to lead to restricted root growth and crop performance.

In many cases, excess water may be removed from the soil through both sensible management of soil structure and appropriate tillage practices. Frequently, however, installation of artificial drainage is necessary. The aim of any artificial drainage programme is to lower the water table so that the amount of plant-available water in the rhizosphere is maximised. If the water table is lowered too much, the soil may become drought-susceptible.

Conversely, if the water table remains too high, waterlogging and poor aeration are likely to continue. The efficiency of drainage depends largely on the lateral spacing, depth and size of drains. Generally, efficiency improves with decreased lateral spacing, although at spacings of greater than around 6 m the rate of inflow to the drains is independent of spacing. Drain depth controls the height of the water table and usually occurs at about 50 cm to 2 m. In terms of size, the ability of a drain to carry water increases with increasing cross-sectional area.

The choice of drainage system depends on a number of factors, the most important being economic controls (installation and maintenance costs, expected yield improvements and availability of subsidies), management considerations (crop and tillage requirements), soil characteristics (texture, hydraulic conductivity and type of drainage problem) and climatic controls (rainfall amount, intensity and seasonality). There are a number of specific drainage techniques which can be employed, either singly or in combination, depending on the type and extent of the drainage problem. Those most commonly used include subsoiling, mole drainage, pipe or tile drainage, open ditches or dykes, and regional or arterial drainage. Subsoiling is a form of deep ploughing and is often used to break up compacted cultivation pans and indurated layers within the soil.

The critical depth for this practice is about 20–50 cm; if it is shallower, then lifting of the soil may occur and if deeper, consolidation of the soil may result. Subsoiling is most effective when the soil is relatively dry and heavy; in wet conditions, plastic deformation and smearing may occur during passage of the subsoiling implement, and this is likely to have an adverse effect on soil structure.

Mole drainage is produced by drawing a mole plough, which is a bullet-shaped instrument usually less than 20 cm in diameter, through the soil to produce a continuous passage. Mole drains are established at a depth of about 40–60 cm, with a lateral spacing of about 2–5 m. They are not permanent features and are susceptible to collapse and sediment blockage. Consequently, they have to be renewed on a regular basis, usually after a period of about 5–7 years.

During the installation of mole drainage, the moisture status of the soil is of crucial importance. If conditions are too wet, smearing and consolidation may result, while shattering may occur if the soil is too dry. In spite of the temporary nature of mole drainage, it is often adopted in preference to other drainage practices because it is cheap and easy to install. It is also commonly used as a form of secondary drainage in combination with primary tile or ditch drains.

Pipe or tile drainage consists of plastic or ceramic pipes of about 10-30 cm in diameter, and is installed at a depth of 0.5-2.0 m. The pipes usually contain perforations or slits in order to speed up the transfer of water from the soil. Once the pipes have been installed, the ditches are often back-filled with gravel, polystyrene or fuel-ash. This helps to improve the flow of water to the drains, while acting as a filter to remove fine sediments which may otherwise lead to drain blockage.

Open ditches or dykes are usually constructed at field margins and are relatively cheap and easy to maintain. This type of drainage has been used since, settled agriculture began, and in Britain dates back to the Roman period. In some circumstances, drainage may need to be improved at the regional scale by arterial drainage. This can be achieved in a number of ways, including straightening of river channels to increase gradient and improve flow, deepening of river channels by dredging, and pumping of water from agricultural land.

Flood prevention is also an important aspect of arterial drainage. In coastal areas, this is achieved through the construction of barrier systems which form part of sea defence strategies. Similarly, in low-lying floodplain areas, it may involve the construction of raised river banks or levées. Such regional or arterial drainage also has a long history, and large areas of land in many coastal, estuarine and low-lying riverine areas have been reclaimed in this way, such as the Fens, Somerset Levels and Romney Marsh in England.

One of the best-known examples of this type of large-scale reclamation is the Polder scheme in the Netherlands, where large areas were reclaimed from the sea in the low-lying coastal fringes. Since, the Zuyder Zee (Reclamation) Act of 1918, four polders have been reclaimed, creating about 165,000 ha of new land. This involved the construction of large dykes which were fed by lateral drains spaced at 1.6–2.0 km intervals. These drains were supplemented by main drains spaced at 300 m intervals and then a coarse network of field drains spaced at 8–24 m intervals.

Although such large-scale drainage programmes have produced some of the world's most fertile and productive land, through improvements to a range of soil physical and chemical characteristics, they have also created problems. Perhaps the most notable of these is soil shrinkage, which is particularly common in areas dominated by marine clays and peats.

In parts of the Fens of eastern England, for example, it is estimated that the surface has been lowered by as much as 5 m following the long history of reclamation and drainage. In an area which is already low lying, surface lowering places even greater pressure on existing flood prevention and drainage schemes, many of which need to be upgraded. Also of concern in areas where large-scale reclamation and drainage are widespread, is the destruction of important wetland habitats, such as those in the Somerset Levels of southwest England.

Soil water deficit occurs most commonly in areas where rainfall is low in comparison with potential evapotranspiration. It is also common in coarse-textured, sandy soils which are excessively drained and low in organic matter, and in fine-textured, clay soils which contain large vertical cracks or which are poorly structured and therefore of low permeability; such soils have a low plant-available water capacity.

Soil water deficit does not necessarily occur throughout the year but is most common during warmer and drier periods. In Britain, for example, it rarely occurs in the October-March period but is common in the April-September period, especially in southern and eastern areas where it may occur in three to five years out of ten. Soil water deficit has the greatest impact where it occurs during the main period of crop growth, when it is demonstrated by excessive wilting of plants.

In some circumstances, intense evaporation may lead to salinisation and sodification of soils. The main way in which soil water deficit is managed is by irrigation. This has been practised since, the beginning of settled agriculture, not only in areas where rainfall is low throughout the year, but also where seasonal drought is a problem, in spite of respectable annual rainfall totals. It is used to replenish the soil moisture store and to leach toxic salts out of the rhizosphere, and as such it is a fundamental component of agriculture in many dryland areas.

The effectiveness of irrigation depends on the method and timing of water application, and on the amount and quality of water applied. There are three main methods—surface application, overhead application and subsurface application.

Surface application usually involves flood and furrow irrigation where water is diverted from existing river channels or canals directly onto the ridge and furrow systems of agricultural fields. This low cost and low technology approach is often preferred in developing countries, although it is also widely used in the developed world. One of the main problems with this method, however, is that capillary action may lead to redistribution of salts from the wet furrow areas to the drier ridges where the crops are often planted.

Overhead irrigation methods include spray or sprinkler, and trickle or drip systems. Spray irrigation systems involve the high pressure application of water through upright nozzles which may be attached to fixed or moveable pipes in a variety of ways, such as self-propelled water guns and centre-pivot systems. Although water can be applied uniformly and efficiently using these methods, equipment costs may be prohibitive and evaporative losses are high, particularly on windy days. Moreover, soil erosion may occur as a result of overhead application, particularly in the early stages of crop growth when vegetation cover is low.

Trickle irrigation involves the application of water through flow regulating emitters which are placed at regular intervals in close proximity to the plants. In this way water is supplied more directly to the plant and evaporative losses are minimised. Perhaps the most effective methods of irrigation involve subsurface applications of water. Here, ditches and underground drains or pipes are used and, although their installation can be costly, evaporation, soil structural damage and erosion losses are minimised.

In addition to irrigation, soil water deficit may also be managed through the control of evaporation from the soil surface. This is most commonly achieved through the addition of organic mulches, or through the establishment of a vegetation cover which can offer some degree of shade. These strategies are usually combined with sensible irrigation practices in order to increase water use efficiency in dry environments.

## DESERT SOILS

A large part of the arid region belonging to the western Rajasthan, Haryana, Punjab, lying between the Indus river and the Aravalli range is affected by the desert conditions of geologically recent origin. This part is covered under a mantle of blown sand which, combined with the arid climate results in poor soil development. The most predominant component of the desert sand is quartz in well-rounded grains but feldspar and hornblende grains also occur with a fair proportion of calcareous grains.

The desert proper, owing to the physiographic conditions of its situation, though lying in the track of the south westerly monsoon receives little rain. The sands which cover the area are partly derived from the disintegration of the subjacent rocks, but are largely blown in from the coastal regions and the Indus Valley, some of these soils contain high percentages of soluble salts, high pH, low loss on ignition, a varying percentage of calcium carbonate and are poor in organic matter.

The Rajasthan desert is a vast sandy plain, including isolated hills or out crops at places. Though on the whole the tract is sandy, the soil improves in fertility from west and north-west to east and north-east. In many parts, the soils are saline or alkaline, with unfavourable physical conditions and high pH.

**Table. The Classification of the Soils According to Soil Taxonomy Along with Traditional Nomenclature.**

| Major group | Soil taxonomy orders |
|---|---|
| Alluvial soil | Entisol<br>Inceptisol<br>Alfisol |
| Black soil | Vertisol<br>Inceptisol<br>Entisol |
| Red soil | Alfisol<br>Inceptisol<br>Ultisol |
| Laterite soil | Alfisol<br>Ultisol<br>Oxisol |
| Desert soil | Entisol<br>Aridisol |

*Problem Soil:* The problem soil are those which owing to land or soil characteristics, cannot be economically used for the cultivation of crops without adoptind proper reclamation measures. Highly eroded soils (sheet and gully), ravine lands, soils on steeply sloping lands, etc., constitute one set of problem soils.

The shallow soil depth, deep gullies, steep and complex slopes are some of the problems which require to be tackled in such areas. Their reclamation may involve massive earth moving operations, terracing, afforestation or plantation to maintain permanent cover with grasses, depending upon the intensity of the problem and the nature of the terrain and soil conditions. The potentiality of the lands, present lands use, cost of operations and other socio-economic factors of the regions are some of the factors which have to be taken into consideration. Acid saline and alkali soils constitute another set of problem soils in the case of which acidity, soluble salts and exchangeable sodium limit the scope of cultivation.

*Acid Soils:* Although soils having pH below 7 are considered to be acidic from the practical standpoint, those with pH less than 5.5 and which respond to liming may be considered to qualify to be designated as acid soils. In the classification of soils both the percentage base saturation and the pH are used as criteria to distinguish acid soils from non-acid ones.

Acid soils occur widely in the Himalayan region, the great eastern plains of extra-peninsular India, the peripheral peninsula and the coastal plains, including the Gangetic delta. They are found to occur on different geological formations under varying physiographical, climatic and vegetational environment. In all these regions the rainfall component of the climate appears to have a dominating influence on the formation of acid soils.

In humid regions, where rainfall is high, the soluble bases formed in the course of weathering of rocks are leached down and carried away by the drainage waters. The continued leaching of soils results in the replacement of calcium, magnesium, potassium and sodium ions by hydrogen ions and the formation of acid soils with low pH. In acid soils the dissolution of aluminosilicate minerals occur and the aluminium ions thus released increase the acidity owing to hydrolysis. Similiarly, humus and hydrous oxides contribute to soil acidity at low pH. Soil acidity exceeding a particular limit is injurious to plant growth. The availability of certain nutrients, particularly phosphorus, calcium and magnesium becomes low with incresing acidity. In acid soils ions of aluminium, iron, manganese and copper may be found in the dissolved form in quantities sufficient to become toxic. Similarly most of the desirable soil microbiological processes such as the activities of the Azotobacter and nodule forming bacteria of legumes are adversely affected as the acidity increases. The satisfactory granulation of soils also becomes difficult to achieve. It is therefore necessary to correct soil acidity for cultivating such soils profitably.

**On the Basis of pH Measurement the Degree of Soil Acidity.**

| pH | Reaction |
|---|---|
| 6.6 to 7.5 | Nearly neutral |
| 6.1 to 6.5 | Slightly acidic |
| 5.6 to 6.0 | Medium acid |
| 5.1 to 5.5 | Strongly acidic |
| 4.6 to 5.0 | Very strongly acid |
| 4.5 and below | Extremely acid |

The pH value of the soil indicates only the active acidity. For remedial measures the total acidity should be considered which is briefly discussed below:

*Lime Requirement:* Acidity in the soil system can be conveniently classified as active or potential acidity. Active acidity includes hydrogen ions in the solution phase and is determined by pH measurements. The potential acidity may be considered to be the exchange acidity and makes up the bulk of the total acidity which is many times greater than the active acidity. In the soil an equilibrium exists between the active and potential acidity.

The lime requirement of a soil necessary to neutralise the total acidity may be defined as the amount of liming material that must be added to raise the pH to some prescribed value. This value is usually in the range of 6 to 7, easily attainable within the optimum growth range of most crops, the pH measurements are widely used to estimate the lime requirements.

The basis for this is that in acid soils a relationship exists between the pH and the percentage base saturation of the soil. Once this relationship is known, it is useful for estimating the lime requirement in the laboratory. However with this "lime requirement" under crops conditions the predicted pH is not generally attained. Hence, a "liming factor" of 1.5 to 2 is commonly used to achieve the desired results, *i.e.,* the lime requirement, as determined in the laboratory, is multiplied by a factor of 1.5 to 2. The approximate amount of limestone required to raise the pH to the neutral level for some soils is given in table.

The quantity of finely ground limestone required to raise the pH value of the ploughed layer (about 15–18 cm) of the soil from the value indicated to a pH of 6.5.

The liming material should be evenly broadcast and worked into the soil several weeks before a crop to allow time for completing the reaction. Keeping the land mois hastens the exchange process. Although liming once in 5 years may serve the purpose the frequency of liming should be determined by making periodic pH measurements.

Table gives only a general idea of the lime requirement on the basis of pH and texture. In several states of India, lime requirement have been worked for specific soils. Hence, these results should be checked by making a reference to the department of agriculture of the state.

*Acid Tolerance of Crops:* Many of the principal crops and vegetables are sensitive to acid soils and suffer injury when grown on them. The optimum pH ranges of some of the crops is given in Table.

This information should be helpful in estimating the lime requirements for a particular crop.

*Saline and Alkali or Sodic Soils:* In many arid and semi arid areas of India crop production is limited because of salinity or alkalinity or both.

It is estimated that about 7 million hectares in the country have either gone out of cultivation or this area produces low yields of crops. The area in different states is given in table.

**The extent of Saline and Alkali Soils in India.**

| State | Area under saline/alkaline soils (million ha) |
|---|---|
| Uttar Pradesh | 1.280 |
| Gujarat | 1.200 |
| West Bengal | 0.840 |
| Rajasthan | 0.720 |
| Punjab | 0.680 |
| Maharashtra | 0.528 |
| Haryana | 0.520 |
| Karnataka | 0.400 |
| Orissa | 0.400 |
| Madhya Pradesh | 0.240 |
| Andhra Pradesh | 0.024 |
| Delhi | 0.016 |
| Kerala | 0.016 |
| Bihar | 0.004 |
| Tamil Nadu | 6.872 (or about 7 million hectares) |

*Classes of Saline and Alkali Soils:* Three classes of saline and alkali soils are recognised:

- *Saline Soils*: The soils containing toxic concentrations of soluble salts in the root zone are called saline soils. Electrical conductivity in the saturation extract of such soils taken as a measure of salts is greater than 4.0 mmhos/cm. Exchangeable sodium percentage is less than 15 and pH is less than 8.5. The soluble salts mainly consist of chlorides and sulphates of sodium, calcium and magnesium. Because of the white encrustation due to salts, the soil is called white alkali.
- *Nin saline alkali or sodic soils*: These soils do not contain any large amount of neutral salts, and the electrical conductivity is less than 4 mmhos/cm. The detrimental effect of alkali soil on plants is largely due to toxicity of a high amount of exchangeable sodium and the pH. Alkali soils have an exchangeable sodium of more than 15 per cent and pH more than 8.5. Such soils have low infiltration rates and the physical condition is unfavourable. Because of high alkalinity, resulting from sodium carbonate, the surface soil is discoloured and black, and hence, it is often called the black alkali.
- *Saline-alkali soils*: This group of soils is both saline and alkali. They have appreciable amounts of suloble salts as indicated by the electrical conductivity values of more than 4 mmhos/cm. Also the exchangeable sodium percentage is more than 1.5. The pH is less than 8.5.

The soil salinity or alkalinity or both have many adverse effects, summarised below:

- Causing low yield of crops or crop failure in extreme cases.
- The limiting of the choice of crops because some crops are sensitive to salinity or alkalinity or both.
- Rendering the quality of fodder as poor, as at times the fodder grown on alkali soils may contain a high amount of molybdenum and a low amount of zinc, causing nutritional imbalance and diseases among live-stock.
- Creating difficulties in the construction of buildings and roads and their maintenance.
- Causing excessive run offs and floods due to low infiltration resulting in damage to crops.

*Salinisation:* Rapid changes in temperature bring about quick weathering of primary minerals and the resulting the ingredients are leached away to the low lying areas. Water is a chief agent by which salts are moved and under hot arid conditions, its evaporation gives rise to accumulation of salts in the soil.

Due to upward and downward movement of the soil solution, the salts gets distributed and accumulate either in the surface soil or in the sub soil and during dry periods, the surface of the soil is covered with salt crust. This is the chief reason for the formation of salt lands. The soil which contains a high concentration of soluble salt is called as saline soil. In most of the saline soils sodium sulphate has been reported to be the chief constituent amounting to 50–80 per cent of the salts forming a crust over the soil. All the Indian soils has been observed to contain NaCl, the quantity of NaCl being more in saline soils.

Soluble salts of Sodium, calcium and magnesium, namely chlorides and sulphates, are the main components present in the salts that lead to the formation of a saline soil. Lack of drainage, nearness of water table within the reach of capillary action and use of irrigation water saline in character, help to accumulate the salts in the profile. Beyond the certain concentration, the soluble salts are always harmful.

According to Magistad, any soil containing more than 0.2 per cent soluble salts can be considered as a saline soil. While the recent classification is based on the conductivity of the saturation extract of the soil which is more than four millions per cm. which corresponds to 0.1 per cent of soluble salts in the case of medium textured soil. The process responsible for the formation of saline soils is called as salinisation.

*Characteristics of Saline Soils:* Saline soils normally do not show any change in structure down the profile but are low in humus and the pH is usually below 8.5. A saline soil may contain over 100 tons/acres of salts in the top 4 feet of soil, they are called as Aolonchack by Russians and white alkali by Americans.

*Harmful Effects of Salinity:* Soluble salts can have two types of effect on the growing plants. Specific effects which are due to particular ions being harmful to the crop can be operative at low or high concentration. The general effects of salinity are due to rise in the osmotic pressure of the soil solution around the roots of the crop and this result in dwarf and stunned growth in the plant.Besides these effects, high concentration of salts affects absorption of water and its translocation in the plant system, root activity and the activity of soil micro-organisms. Amongst the three sodium salts, sodium carbonate is extremely toxic, sodium chloride is intermediate in its effect, while sodium sulphate is least toxic (Hilgard). The salinity of the surfaces soil is predominently governed by the annual rainfall, while the average salinity of the whole profile is determined by annual temperature.

*Alkalisation:* Alkalisation is a next step to salinisation. It takes place due to preponderance of Na ions in the exchange complex. When Na from the soluble salts enters into clay complex (due to base exchange reactions) and its proportion increases, the soils becomes alkaline and the entry of Na ion on the exchange complex Chopan character. The process of alkalisation can be represented schematically as follows:

Clay complex + Soluble Na salts $\rightarrow$ alkali soil + Soluble salts of other bases

*Characteristics of Alkali Soils:* Alkali soils are termed as black alkali by Americans ans solonetz by Russians. Such soils contain free $N_2CO_3$ and are alkaline in reaction. Exchangeable Na is more than 15 per cent of the exchangeable bases. The soils are defloculated and impervious. pH is usually very high and the soils show swelling capacity.

Saline nature of the soil, high concentration of sodium salts, basin shape topography, faulty drainage and arid nature of the climate accelerate the process of alkalisation. A saline soil contains excess quantity of soluble salts, while alkali soil contains excess quantity of exchangeable Na. A soil which contains an excess of both sodium and soluble salts is termed as saline-alkali soil.

*Harmful Effects of Alkalinity:* Harmful effects of high alkalinity are manifested through non-availability of nutrient elements like iron, manganese, boron and phosphate. High pH of alkali soils is due to increase in the concentration of exchange. Na, which in turn modifies the physical conditions of the soil and makes the soil sticky and impermeable and brings out wilting and scorching effect on plants.

*Salts Index of Irrigation Water:* With irrigation, huge quantities of soluble salts are added to the soil. To avoid the accumulation of salts, the output of salt through drainage should be more than the input of salt through irrigation. Puri has suggested a formula for determining the salt index of irrigation water. Salt index is negative for good water and positive for those water that are unsuitable for irrigation purposes.

Generally, it may be said that waters with a T. S. S. (Total Soluble Salts) content below 60 parts per 100,000 parts are suitable for irrigation those with salt content between 60 to 120 are suitable if the salt index is negative and waters with a salt content above 120 are unsuitable even though the salt index is negative.

According to the standards prescribed by the U. S. D. A. salinity Laboratory, the quality of irrigation water is decided by three factors namely-Total salt content, Soluble sodium per cent and the boron content. The electrical conductivity is used as an index for the total salt content of irrigation water and based on this value, the water is classified into three categories as given below-

The soluble Na per cent of water forms an important index of its quality and suitability for irrigation. The solube Na per cent should be less than 60 to consider water suitable for irrigation purpose. Based upon the tolerance and sensitivity of different crops, and based upon the Boron contents of irrigation water, it is classified into three categories.

Recently the quality of irrigation water is also judged or assessed on the basis of number of c. cs of E. D. T. A. reagent consumed by 5 ml. of water.

*Movement of the Salt in the Soil profile:* salts can be move in the soil both in the vertical as well as in the lateral direction. When a constant head of water is maintained, Chandnani observed that in a sandy loam soil from sindh, the average movement of the salt was one foot per month in the lateral direction, but after a period of six months, the lateral movement of salt was rather little. The same author further observed that of the injurious salts, the more injurious NaCl was washed down more quickly than the less injurious $Na_2SO_4$. Salts once washed down come up to the surface (depending on the capillary activity and surface evaporation), the more injurious NaCl comes up more quickly than the less injurious $Na_2SO_4$. Cropping retards this rise of salt in the profile.

*Pattern of Distribution of Salts in the Profile:* With low rainfall and high temperature, salts try to accumulate at the surface layer of arid region soils, while they go down in humid soils, with high rainfall. On the basis of the nature and concentration of soluble salts at different depths, Indian soils can be classifieds into five main groups.

- *Group I*: Characterised by the predominance of NaCl Concentration of soluble salt highest at the surface and lowest at the bottom.
- *Group II*: Sulphate very low. Soluble salts highest at the top but not lowest at the bottom.
- *Group III*: Characterised by the predominance of either Ca, Na, $SO_4$ and Cl, concentration of salt is highest neither at the surface nor at the bottom.
- *Group IV*: Characterised by very high salt content, the concentration being highest at the bottom.

- *Group V*: Characterised by low content of soluble salts which are distributed evenly throughout the profile.

*Adverse Effects of Salinity on Soil Fertility:* The adverse effect of salinity on soil fertility are numerous. Its effects on the uptake of nutrients, absorption of moisture and soil structure are well known. Nitrogen fixing organisms have been known to be remarkably resistant to the action of high concentration of soluble salts in the soil. Limpan and Sharp found that NaCl was not toxic to Azotobacter upto 0.5 per cent concentration.

Tamhane and Krishna observed that appreciably large quantities of salt in kallar soils of sindha amounting to 1.5 per cent only rendered the organism slightly inactive, while Iswaran and Sen from their laboratory studies in normal soils soncluded that fixing of atmospheric N by Azotobacter was adversely affected by NaCl, and the concentration of NaCl in the soil at which fixation of atmospheric N by Azotobacter was inhabited varied from 1.3 per cent in Sommagaon (Assam) soil to 4.32 per cent in Naghs (Ajmer) soil.

During the process of reclamination, the soil losses very important neutrient elements like mangnese, nitrogen and phosphate. Tomaintain fertility in such soils, it would be necessary to manure the soil with a legume which would supply requisite amount of these neutrients. Saline soils rich in Nacl have been found to be devoid of $Na_2 > N$, while those containing $SO_4$ contain appreciable quantities.

*Evaluation of Saline Soils:* The electrical conductivity measurement affords a rapid and accurate method for determining the amount of soluble salts in the soil. The conductivity increases with the increasing amount of salts in the soil. The effect of salinity on crop growth can be well assessed from the date given in table.

The characteristic of saline and saline alkali soils in of conductivity, pH and exchangeable Na are illustrated in table.

Evaluation of saline and alkali soils is a difficult task. But while doing so, importance should be given to osmotic pressure, electrical conductivity, soil pH and gypsum requirements. The osmotic pressure bears a correlation with electrical conductivity and it's expressed by the following equation.

$$\text{O.P.} = (\text{E.C.}) * (0.36).$$

The amount of gypsum required to replace exchangeable Na by exchangeable Ca can be worked out from the data of Richards reproduced in Table. The relation between an amendment such as gyo\psum and exchangeable Na is an equilibrium reaction, and therefore it does not go entirely to completion. To appraise the effect of soluble salts the soil on crop growth, Scofield proposed a salinity scale.

This scale does not take into consideration the effect of soil pH on crop growth. Agrawal and Yadav studied the combined effect of soil pH and salinity and have suggested a new scale to appraise the effect of soluble salts on crop growth. These workers consider that pH value, Electrical conductivity of saturation extract, cation exchange capacity using Na-acetate and the degree of calcareousness of the sample should from a workable criteria for salinity and alkali appraisal of the Indian soils.

## MAJOR ELEMENTS OF SOIL

Eight chemical elements comprise the majority of the mineral matter in soils. Of these eight elements, oxygen, a negatively-charged ion (anion) in crystal structures, is the most prevalent on both a weight and volume basis. The next most common elements, all positively-charged ions (cations), in decreasing order are silicon, aluminum, iron, magnesium, calcium, sodium, and potassium. Ions of these elements combine in various ratios to form different minerals. More than eighty other elements also occur in soils and the earth's crust, but in much smaller quantities.

Soils are chemically different from the rocks and minerals from which they are formed in that soils contain less of the water soluble weathering products, calcium, magnesium, sodium, and potassium, and more of the relatively insoluble elements such as iron and aluminum. Old, highly weathered soils normally have high concentrations of aluminum and iron oxides.

The organic fraction of a soil, although usually representing much less than 10 per cent of the soil mass by weight, has a great influence on soil chemical properties. Soil organic matter is composed chiefly of carbon, hydrogen, oxygen, nitrogen and smaller quantities of sulphur and other elements. The organic fraction serves as a reservoir for the plant essential nutrients, nitrogen, phosphorus, and sulphur, increases soil water holding and cation exchange capacities, and enhances soil aggregation and structure.

The most chemically active fraction of soils consists of colloidal clays and organic matter. Colloidal particles are so small ($< 0.0002$ mm) that they remain suspended in water and exhibit a very large surface area per unit weight. These materials also generally exhibit net negative charge and high adsorptive capacity. Several different silicate clay minerals exist in soils, but all have a layered structure.

Montmorillonite, vermiculite, and micaceous clays are examples of 2:1 clays, while kaolinite is a 1:1 clay mineral. Clays having a layer of aluminum oxide (octahedral sheet) sandwiched between two layers of silicon oxide (tetrahedral sheets) are called 2:1 clays. Clays having one tetrahedral sheet bonded to one octahedral sheet are termed 1:1 clays.

### CATION EXCHANGE

Silicate clays and organic matter typically possess net negative charge because of cation substitutions in the crystalline structures of clay and the loss of hydrogen cations from functional groups of organic matter. Positively-charged cations are attracted to these negatively-charged particles, just as opposite poles of magnets attract one another. Cation exchange is the ability of soil clays and organic matter to adsorb and exchange cations with those in soil solution (water in soil pore space).

A dynamic equilibrium exists between adsorbed cations and those in soil solution. Cation adsorption is reversible if other cations in soil solution are sufficiently concentrated to displace those attracted to the negative charge on clay and organic matter surfaces. The quantity of cation exchange is measured per unit of soil weight and is termed cation exchange capacity.

Organic colloids exhibit much greater cation exchange capacity than silicate clays. Various clays also exhibit different exchange capacities. Thus, cation exchange capacity of soils is dependent upon both organic matter content and content and type of silicate clays.

*Cation exchange capacity is an important phenomenon for two reasons:*

1. Exchangeable cations such as calcium, magnesium, and potassium are readily available for plant uptake and
2. Cations adsorbed to exchange sites are more resistant to leaching, or downward movement in soils with water.

Movement of cations below the rooting depth of plants is associated with weathering of soils. Greater cation exchange capacities help decrease these losses. Pesticides or organics with positively charged functional groups are also attracted to cation exchange sites and may be removed from the soil solution, making them less subject to loss and potential pollution.

Calcium ($Ca^{++}$) is normally the predominant exchangeable cation in soils, even in acid, weathered soils. In highly weathered soils, such as oxisols, aluminum ($Al^{+3}$) may become the dominant exchangeable cation.

The energy of retention of cations on negatively charged exchange sites varies with the particular cation. The order of retention is:

aluminum > calcium > magnesium > potassium > sodium > hydrogen.

Cations with increasing positive charge and decreasing hydrated size are most tightly held. Calcium ions, for example, can rather easily replace sodium ions from exchange sites.

This difference in replaceability is the basis for the application of gypsum ($CaSO_4$) to reclaim sodic soils (those with > 15 per cent of the cation exchange capacity occupied by sodium ions). Sodic soils exhibit poor structural characteristics and low infiltration of water. The cations of calcium, magnesium, potassium, and sodium produce an alkaline reaction in water and are termed bases or basic cations.

Aluminum and hydrogen ions produce acidity in water and are called acidic cations. The percentage of the cation exchange capacity occupied by basic cations is called per cent base saturation. The greater the per cent base saturation, the higher the soil pH.

## SOIL PH

Soil pH is probably the most commonly measured soil chemical property and is also one of the more informative. Like the temperature of the human body, soil pH implies certain characteristics that might be associated with a soil. Since, pH (the negative log of the hydrogen ion activity in solution) is an inverse, or negative, function, soil pH decreases as hydrogen ion, or acidity, increases in soil solution. Soil pH increases as acidity decreases.

A soil pH of 7 is considered neutral. Soil pH values greater than 7 signify alkaline conditions, whereas those with values less than 7 indicate acidic conditions. Soil pH typically ranges from 4 to 8.5, but can be as low as 2 in materials associated with pyrite oxidation and acid mine drainage. In comparison, the pH of a typical cola soft drink is about 3. Soil pH has a profound influence on plant growth. Soil pH affects the quantity, activity, and types of microorganisms in soils which in turn influence decomposition of crop residues, manures, sludges and other organics.

It also affects other nutrient transformations and the solubility, or plant availability, of many plant essential nutrients. Phosphorus, for example, is most available in slightly acid to slightly alkaline soils, while all essential micronutrients, except molybdenum, become more available with decreasing pH. Aluminum, manganese, and even iron can become sufficiently soluble at pH $< 5.5$ to become toxic to plants. Bacteria which are important mediators of numerous nutrient transformation mechanisms in soils generally tend to be most active in slightly acid to alkaline conditions.

## CONCLUSION

Soils widely vary in their characteristics and properties. In order to establish the interrelationship between their characteristics they require to be classified. Understanding the properties of the soils is important in respect of the optimum use they can be put to and for their best management requirements. Classification helps to reduce the study of the number of individuals to a few well defined units. It helps to develop a legend for mapping soils during the surveys. It helps to group together such soils as have comparable characteristics so that the knowledge regarding them is presented in a systematic manner.

To classify soils and group them together in a meaningful manner different systems of soil classification have been used from time to time, these systems have varied over a period of time having been drawn up to meet the requirements and the immediate purposes of their use. As a knowledge of the soils which helps one to understand their genesis has developed the systems of classification have also developed.

The modern system of classification "Soil Taxonomy" developed by the USDA has been recommended for adoption all over the world and in this country as a result of the decision taken at the All-India Workshop held in 1969. This is a multi category system which has six categories: order, sub-order, great group, sub-group, family and series. The system id heirarchical in that an order is divided into sub-orders and sub-orders into great groups, great groups into sub groups, etc., upto the series level. Thus the number in the higher categories is fixed whereas in the lower categories it varies. The soil series is the basic unit of classification and in all our surveys the soil series are first identified and described.

Whereas phases of soil series are used as mapping units in detailed soil surveys the associations of soil series are used as mapping units in reconnaissance soil surveys. The progressive abstraction and compilation of reconnaissance maps help one to construct small scale maps at the state and country levels with the associations of higher categories as the mapping units. Thus, for example, great groups, or the association of great groups are the units delineated at the state or country level.

Brief descriptions of the major soil groups along with their modern nomenclature are as follows:

*Alluvial Soils:* These include the deltaic alluvium, calcareous alluvial soils, coastal alluvium, and coastal sands. This is by far the largest and most important soil group of India contributing the largest share to the agricultural wealth. In this immense tract though a great deal of variation exists, the main features of the soil are derived from the deposition laid by the numerous tributaries of the Indus, the Ganges, and the Brahmaputra systems. These streams, draining the Himalayas, bring with them the products of weathering of rocks constituting the mountains in various degrees of fineness and deposit them on the plains.

Geologically, the alluvium is divided into khadar and bhangar. The soils differ from the drift sands to loams and from fine silts to stiff clays. Impervious clays obstruct the drainage and promote the accumulation of injurious salts of sodium and magnesium and make the soils sterile.

The formation of hard-pans at certain levels in the soil profile due to binding of soil grains by silica or calcerius matter forming an impervious layer is often observed in these alluvial soils. Layers of kankars in the Indo-gangetic alluvium of Uttar Pradesh and West Bengal and alsi occasionally layers composed of impure iron oxides are instances of the formation of hard pans.

The most important characteristic of the soil of Assam is its acidity. Generally, those on the old alluvium and hills are more acidic than the new alluvial soils along the river banks which are neutral or alkaline.

In West Bengal, the portions Murshidabad, Bankura the whole of Burdwan and the western half of Midnapore comprising the rarh region, are composed mainly of the old alluvium. There is hardly any regularity in the deposition of river-borne materials. Deposits laid down earlier, subjected to climate and other influences have led to soils varying from one another in colour, texture, chemical, physical and other properties.

The alluvial soils of Bihar may be divided into 2 main divisions:

1. The alluvium north of the ganges;
2. The alluvium south of the ganges.
   - The northern alluvium comprises the area between the Himalayas in the north and the ganges in the south. The soil is alluvial with a calcerous belt in the form of a triangle in the west and inundated areas in the middle; these areas remain flooded for different periods in the year. The soils are sandy loams to clayeye loams and neutral to alkaline. Their CaO ranges from 0.5 to 20 per cent. They are rich in total and available potash but deficint in phosphorus.
   - The southern alluvium comprises the area between the Ganges in the north and the hilly region in the south. The soils vary in colour and texture from light greyish loams to heavy black clays. The area has been divided into:
   - Highland soils;
   - Soils liable to inundation;
   - Saline soils,
   - Diara land soils.

The soils of Uttar Pradesh are divided into four classes:

- The alluvium of the west and north-west, lighter in texture,
- The alluvium in the centre, intermediate between light and heavy texture,
- The alluvium in the north-east developed on a calcerous parent material.

The soils contain varying amounts of $CaCO_3$ and soluble salts and are neutral to alkaline. The lime content increases at lower depths. They are generally poor in $P_2O_5$, nitrogen and organic matter.

On the coast of Orissa, there are stretches of sand and sand-hills alternating with deltaic swamps. Behind this is an area of cultivated alluvium. Soils are both sandy and fine; with enough potash but not enough $P_2O_s$.

The alluvial soils of Tamil Nadu are found in the delataic regions and along the coasts. The profile reveals alternate layers of sand and silt. The composition of the strata varies with the nature of the silt brough in by the rivers.

The stratified deposition is confined to ares very near the river courses. But away from the rivers the soils are heavy throughout, the texture ranging from clay loam to heavy clay through silty clay. Sandy layers occur at very low depths. In the Gujarat state, the alluvial soils are confined to the northern Gujarat tract, Ahmedabad and Kaira districts and are locally known as goradu. The gorat soil of Baroda corresponds to the older alluvium consisting of brown clay with kankar. Those from the recent depositions are known as bhata. The light-sandy, red and yellow soils found in the Mahanadi basin (Madhya Pradesh) including the Balaghat and three districts of Durg, Raipur and Bilaspur are of alluvial origin.

The soils of the Punjab and Haryana plains belong to the same class as the Idogangetic plain. They are loams or sandy loams consisting of a soil crust of varying depth, with soluble salts in amounts. The lower layer contains kankar nodules. Due the presence of sodium, the soils are alkaline, with adequate phosphorus and potash, deficient in organic matter and nitrogen.

In kerala, there are 2 types of alluvial soils *viz.* the coastal alluvium and the alluvium. In central Kerala, the width of the coastal alluvial tracts increases, in the north and south they are comparatively narrow. The alluvial soils of Kuttanad form a low lying area, once a part of the sea, later filled up by the silt carried down by the pampa and other rivers. The coastal alluviums are sandy with low water retention and low nutrient status. The alluviums on the river banks are fertile.

*Black Soils:* These vary in depth from shallow to deep. The typical soil derived from the deccan trap is the regur or black cotton soil. It is common in Maharashtra, western parts of Madhya Pradesh, parts of Andhra Pradesh, parts of Gujarat and some parts of Tamil Nadu. It is comparable with the 'chernozems' of Russia and the 'prairie soil' of the United States. It is derived from 2 types of rocks, the Deccan and the Rajmahal trap and forriginous gneisses and schists occurring in the Tamil Nadu state under semi-arid conditions. There is no change in colour upto a thickness of 2 to 3 metres.

Many black soil areas have a high degree of fertility but some, in the uplands, are rather poor. In the broken country, in the hills and plains, they are darker, deeper and richer, and are being constantly enriched by the additions washed down the hills.

Black soils are height argillaceous, fine grained and dark and contain a high proportion of calcium and magnesium carbonates. They are sticky when wet. Due to contraction on drying, large and deep cracks are formed. They contain abundant iron, lime magnesia and alumina. They are poor in phosphorus, nitrogen and organic matter in all regur areas, and in those derived from ferro-magnesium schists, in particular, there is a layer rich in kankar nodules formed by the ssegregation of calcium carbonate. The soils are generally rich in montmorillonitic and beidellitic group of clay minerals.

In maharashtra, the deccan trap occupies a large area. On the uplands and slopes, the soils are thin and poor. On the lowlands and in the valleys, deep and relatively clayey black soils are found. Along the ghats, the soils are very coarse and gravelly. In the valleys of the Tapti, narmada, the Godavari and the Krishna rivers, heavy black soil is often 6 metres deep. The subsoil contains a good amount of lime. The black cotton soil dominates the Surat and Broach districts. Degraded solonised black soils locally known as chopan occur in the canal zones of the Deccan in maharashtra.

In Tamil Nadu, the black soils are either deep or shallow and may contain gypsum. Soils are fine textured, have high pH (8.5-9) and are rich in lime (5-7 per cent) They have low permeability and high hygroscopic coefficient, pore space, maximum water holding capacity and and true specific gravity. Black soils have a high base status and a high cation exchange capacity of 40 to 60 m.e.per 100 g. The iron content varies from 10 to 13 per cent and the $SiO_2/R_2O_3$ varies from 3 to 3.5, CaO and MgO contents are high.

In Madhya Pradesh, 2 distinct kinds of black soils are found, *viz.*:

1. Deep heavy black soils covering the Narmada valley,
2. Shallow black soils in other areas.

The cotton growing areas are covered by the deep heavy black soils, but there are also soils of lighter texture. The clay content varies from 35 to 50 per cent and the $SiO_2/R_2O_3$ varies from 3 to 3.5. The organic matter content is low. The black soils of Karnataka are fine-textured with varying salt concentration. The soils are rich in lime and magnesia, the $SiO_2/R_2O_3$ ratio of clay fraction is 3.6.

*Red Soils:* The soils comprise vast areas of Tamil nadu, Karnataka, Goa, Daman and Diu, south eastern Maharashtra, Madhya Pradesh, orissa and Chhotanagpur in the north it includes the Santhal Paraganas in Bihar, the Birbhum district of West Bengal, the Mirzapur, Jhansi and hamirpur district of Uttar Pradesh.

The ancient crystalline and metamorphic rocks on weathering have given rise to the red soils. The red colour is due to the wide diffusion of iron than to the high proportion of it. They are generally poor in nitrogen, phosphorus and humus. These soils are poorer in lime, potash, iron oxide and phosphorus than the regur soils. Many of the so-called red soils of southern India are not red. On the other hand, some red soils are of the lateritic origin and quite different in nature.

The clay fraction of red soils is rich in Kaolinite. Red soils are also found under forest vegetation. Red and yellow soils are also seen side by side. The yellow colour is due to the high degree of hydration of the ferric oxide in them than that in the red soils. Morphologically, the red soils can be divided into 2 sub-groups:

1. Red loams, characterised by argillaceous soils with a clody structure;
2. Red earths where the top soil is rich and friable and rich in secondary concretions.

The red soils in Tamil Nadu occupy nearly two thirds of the cultivated area. They are all in situ formations of the underlying rocks under climatic conditions. The rocks are micaceous or red granites. The soils are shallow, open in texture, pH from 6 to 8, low base status and low exchange capacity.

They are also deficient in organic matter and poor in plant nutrients. $SiO_2/R_2O_3$ ratios of 2.5 to 3.0.

The predominant soil in the eastern tract of karnataka is the red soil overlying the granite from which it is derived. Especially found in the districts of Bangalore, Kolar, Mysore, Tumkur, and Mandya. There are shades from red to yellows. Loamy red soils are predominant in Shimoga, Hassan and Kadur. They are rich in $K_2O$ and contain sufficient $P_2O_5$ (0.05 -0.3 per cent) and lime content of 0.1 to 0.8 per cent. Nitrogen is below 0.1 per cent. Iron and Alumina are high, 30 to 40 per cent. A broad strip of the area from eastern to western Coorg is red loam, easily drained, with a dense growth of trees.

The acid soils towards the south of Bihar, those of Ranchi, Hazaribagh, Santhal Paraganas, Manbhum Singhbhum are red soils. The pH varies from 5 to 6.8. Another distinctive feature is the higher percentage of acid soluble $Fe_2O_3$ than $Al_2O_3$. Available potash is sufficient but $P_2O_5$ is low. In West Bengal the red soils are the transported soils from the Chhotanagpur plateau.

A typical profile of the red sil at Chandkhuri Farm, Raipur, Madhya Pradesh, reveals that the percentage of concretions increases down the profile. The total exchangeable cations are 20 m.e.per 100g. The $SiO_2/R_2O_3$ ratio of the clay fraction is between 2 and 3 and the C/N ratio is approximately ten.

A part of the Jhansi district in Uttar Pradesh comprises red soils. There are 2 types:

- *Parwa*: The parwa is a brownish grey soil varying from good loam to sandy or clay loam.
- *Rakkar*: The rakkar is the true red soil which is generally not useful for cultivation.

The soils of Banaras and Mirzapur developed on the Vindhyan parent materials have also been classified as tropical and subtropical red loams.

In the Telengana Division of Andhra Pradesh where the predonant geological formation is granite and a gheissic complex, both red and black soils predominate. The red soils or chalkas are sandy loams located at higher levels. Such soils are utilised for the cultivation of Kharif crops.

*Laterites and Lateritic Soils:* Laterite is a formation peculiar to India and some other tropical countries with an intermittently moist climate. It is a vesicular rock composed essentially of a mixture of the hydrated oxides of aluminium and iron with small amounts of manganese oxides, titania, etc. It is derived from the atmospheric weathering of several types of rocks. Under the monsoon condirions of alternating wet and dry seasons, the siliceous matter of the rocks is leached away almost completely during weathering.

The laterite may become broken off and be carried to lower levels by streams, and when redeposited may get comented again into a compact mass by the segregative action of the hydrated oxides. Thus there are high level laterites resting on the rocks resting on low level laterites formed in the usual way of detrital deposits. Laterites are specially well developed on the hill of Karnataka, Kerala, Madhya Pradesh, Eastern Ghat region of Orissa, Maharashtra, West Bengal Tamil Nadu and Assam. All lateritic soils are very poor in lime and magnesia and are deficient in nitogen. $P_2O_5$ may be high in the form of iron phosphate, but $K_2O$ is deficient, there is a higher content of humus.

In Tamil Nadu, there are high level and low level laterites formed under peculiar climatic and weather conditions. They are both in situ and sedimentary formations are found all along the West coast where rainfall is heavy and humid climate prevails.

On the laterites, at low elevations, paddy is grown, on high elevations tea, cinchona, rubber and coffee are grown. The soils are rich in nutrients and contain 10–20 per cent organic matter. The pH is low (3.5-4), and the higher the elevation, the more acidic the soil. In the laterite soils of Ratnagiri, the coarse material is found in large quantities. These soils are rich in plant food constituents, except lime.

In Kerala, both high level and low level laterites occur. The high level laterites are rich soils because of their proper management. The laterites on lower elevations have a poor nutrient status. Those of the West coast grow plantation crops, *e.g.,* tea, rubber, coconut, and arecanut, but at low elevations paddy is also grown. The soils are poor in N P K and organic matter, pH ranging from 4.5 to 6.0. $SiO_2/R_2O_3$ ratio increases with the decreasing elevation.

The laterite soils in Karnataka occur in the western parts in the districts of North Kanara and South Kanara, Shimoga, Hassan, Kadur, and Maysore. All the soils are comparable with the laterites and similiar formations in Malabar and Nilgiris district. These soils are very low in bases, because of the severe leaching and erosion. These soils are also poor in $P_2O_5$. Their pH is not so low as that of plantation soils.

In West Bengal the area between the Damodar and Bhagirathi is interspersed with some basaltic and granitic hills, with lateritic capping. In these soils the $SiO_2/Al_2O_3$ ratio of the clay fraction is quite high. The percentages of $K_2O$, $P_2O_5$ and N are low showing considerable leaching and washing out of these substances due to chemical weathering. In Bihar, the lateritic soils occur principally as a cap on the higher plateaus, but is also found in some valleys.

The laterites of Orissa are found largely capping the hills and plateaus occasionally in considerable thickness. Large areas in Khurda are occupied by laterites; those of Balasore are gravelly and appear to be detrital.

# 3

# Water: Risk and Management

## WATER POPULATION GROWTH AND ENVIRONMENTAL QUALITY

There is a long history of study and debate about the interactions between population growth and the environment. According to the British thinker Malthus, for example, a growing population exerts pressure on agricultural land, causing environmental degradation, and forcing the cultivation of land of poorer and poorer quality. This environmental degradation ultimately reduces agricultural yields and food availability, causes famines and diseases and death, thereby reducing the rate of population growth. Population growth, because it can place increased pressure on the assimilative capacity of the environment, is also seen as a major cause of air, water, and solid-waste pollution. The result, Malthus theorized, is an equilibrium population that enjoys low levels of both income and environmental quality. Malthus suggested positive and preventative forced control of human population, along with abolition of poor laws.

Malthus theory, published between 1798 and 1826, has been analyzed and criticized ever since. The American thinker Henry George, for example, observed with his characteristic piquancy in dismissing Malthus: "Both the jayhawk and the man eat chickens; but the more jayhawks, the fewer chickens, while the more men, the more chickens." Similarly, the American economist Julian Lincoln Simoncriticised Malthus's theory. He noted that the facts of human history have proven the predictions of Malthus and of the Neo-Malthusians to be flawed. Massive geometric population growth in the 20th century did not result in a Malthusian catastrophe.

The possible reasons include: increase in human knowledge, rapid increases in productivity, innovation and application of knowledge, general improvements in farming methods (industrial agriculture), mechanization of work (tractors), the introduction of high-yield varieties of wheat and other plants (Green Revolution), the use of pesticides to control crop pests. More recent scholarly articles concede that while there is no question that population growth may contribute to environmental degradation, its effects can be modified by economic growth and modern technology. Research in environmental economics has uncovered a relationship between environmental quality, measured by ambient concentrations of air pollutants and per capita income. This so-called environmental Kuznets curve shows environmental quality worsening up until about $5,000 of per capita income on purchasing parity basis, and improving thereafter. The key requirement, for this to be true, is continued adoption of technology and scientific management of resources, continued increases in productivity in every economic sector, entrepreneurial innovation and economic expansion.

Other data suggests that population density has little correlation to environmental quality and human quality of life. India's population density, in 2011, was about 368 human beings per square kilometer. Many countries with population density similar or higher than India enjoy environmental quality as well as human quality of life far superior than India.

For example: Singapore (7148/km$^2$), Hong Kong-China (6349/km$^2$), South Korea (487/km$^2$), Netherlands (403/km$^2$), Belgium (355/ km$^2$), Japan (337/ km$^2$), England (395/km$^2$), the state of Florida (353/ km$^2$), the state of New York (412/ km$^2$), the state of Massachusetts (840/ km$^2$), the state of North Rhine-Westphalia (523/ km$^2$), the region of Île-de-France (974/ km$^2$), and the region of Lombardy (417/ km$^2$).

## CAUSE OF WATER POLLUTION

A 2007 study finds that discharge of untreated sewage is single most important cause for pollution of surface and ground water in India. There is a large gap between generation and treatment of domestic wastewater in India. The problem is not only that India lacks sufficient treatment capacity but also that the sewage treatment plants that exist do not operate and are not maintained. Majority of the government-owned sewage treatment plants remain closed most of the time due to improper design or poor maintenance or lack of reliable electricity supply to operate the plants, together with absentee employees and poor management. The wastewater generated in these areas normally percolates in the soil or evaporates. The uncollected wastes accumulate in the urban areas cause unhygienic conditions and release pollutants that leaches to surface and groundwater.

A 1992 World Health Organization study is claimed to have reported that out of India's 3,119 towns and cities, just 209 have partial sewage treatment facilities, and only 8 have full wastewater treatment facilities. Downstream, the untreated water is used for drinking, bathing, and washing. A 1995 report claimed 114 Indian cities were dumping untreatedsewage and partially cremated bodies directly into the Ganges River. Open defecation is widespread even in urban areas of India. This situation is typical of India as well as other developing countries.

According to another 2005 report, sewage discharged from cities and towns is the predominant cause of water pollution in India. Investment is needed to bridge the gap between 29000 million litre per day of sewage India generates, and a treatment capacity of mere 6000 million litre per day. A large number of Indian rivers are severely polluted as a result of discharge of domestic sewage.

The Central Pollution Control Board, a Ministry of Environment and Forests Government of India entity, has established a National Water Quality Monitoring Network comprising 1429 monitoring stations in 27 states and 6 in Union Territories on various rivers and water bodies across the country. This effort monitors water quality year round. The monitoring network covers 293 rivers, 94 lakes, 9 tanks, 41 ponds, 8 creeks, 23 canals, 18 drains and 411 wells distributed across India. Water samples are routinely analyzed for 28 parameters including dissolved oxygen, bacteriological and other internationally established parameters for water quality. Additionally 9 trace metals parameters and 28 pesticide residues are analyzed. Biomonitoring is also carried out on specific locations.

The scientific analysis of water samples from 1995 to 2008 indicates that the organic and bacterial contamination are severe in water bodies of India. This is mainly due to discharge of domestic wastewater in untreated form, mostly from the urban centres of India.

In 2008, the water quality monitoring found almost all rivers with high levels of BOD. The worst pollution, in decreasing order, were found in river Markanda (590 mg O/l), followed by river Kali (364), river Amlakhadi (353), Yamuna canal (247), river Yamuna at Delhi (70) and river Betwa (58). For context, a water sample with a 5 day BOD between 1 and 2 mg O/L indicates a very clean water, 3 to 8 mg O/L indicates a moderately clean water, 8 to 20 indicates borderline water, and greater than 20 mg O/L indicates ecologically-unsafe polluted water.

The Mithi River, which flows through the city of Mumbai, is heavily polluted. The levels of BOD are severe near the cities and major towns. In rural parts of India, the river BOD levels were sufficient to support aquatic life.

Total coliform and fecal coliform densities in the rivers of India range between 500 to 100,000 MPN/100 ml. The presence of coliform suggests that the water is being contaminated with the fecal material of humans, livestocks, pets and other animals. Rivers Yamuna,Ganga, Gomti, Ghaggar, Chambal, Mahi, Vardha are amongst the other most coliform polluted water bodies in India. For context, coliform must be below 104 MPN/100 ml, preferably absent from water for it to be considered safe for general human use, and for irrigation where coliform may cause disease outbreak from contaminated-water in agriculture.

In 2006, 47 per cent of water quality monitoring stations in India reported a total average annual coliform levels above 500 MPN/100 ml. During 2008, 33 per cent of all water quality monitoring stations reported a total coliform levels exceeding those levels, suggesting recent effort to add pollution control infrastructure and upgrade treatment plants in India, may be reversing the water pollution trend. Treatment of domestic sewage and subsequent utilization of treated sewage for irrigation can prevent pollution of water bodies, reduce the demand for fresh water in irrigation sector and become a resource for irrigation. Since 2005, Indian wastewater treatment plant market has been growing annually at the rate of 10 to 12 per cent. The United States is the largest supplier of treatment equipment and supplies to India, with 40 per cent market share of new installation. At this rate of expansion, and assuming the government of India continues on its path of reform, major investments in sewage treatment plants and electricity infrastructure development, India will nearly triple its water treatment capacity by 2015, and treatment capacity supply will match India's daily sewage water treatment requirements by about 2020.

Water resources have not been linked to either domestic or international violent conflict as was previously anticipated by some observers. Possible exceptions, notes a 2004 report, include some communal violence related to distribution of water from the Kaveri River and political tensions surrounding actual and potential population displacements by dam projects, particularly on the Narmada River. A 1997 article claimed Punjab is another hotbed of pollution, for example, Buddha Nullah, a rivulet which run through Malwaregion of Punjab, India, and after passing through highly populated Ludhiana district, before draining into Sutlej River, a tributary of the Indus river, is today an important case point in the recent studies, which suggest this as another Bhopal in making.

A joint study by PGIMER and Punjab Pollution Control Board in 2008, revealed that in villages along the Nullah, calcium, magnesium, fluoride, mercury, beta-endosulphan and heptachlor pesticide were more than permissible limit (MPL) in ground and tap waters. Plus the water had high concentration of COD and BOD (chemical and biochemical oxygen demand), ammonia, phosphate, chloride, chromium, arsenic and chlorpyrifos pesticide. The ground water also contains nickel and selenium, while the tap water has high concentration of lead, nickel and cadmium. The Hindon River, which flows through the city of Ghaziabad, highly polluted and groundwater of this city has coloured and poisoned by industrial effluents, Hindon Vahini is strongly opposing of water pollution activities. Flooding during monsoons worsens India's water pollution problem, as it washes and moves all sorts of solid garbage and contaminated soils into its rivers and wetlands.

The annual average precipitation in India is about 4000 billion cubic metres. From this, with the state of Indian infrastructure in 2005, the available water resource through the rivers is about 1869 billion cubic meters. Accounting to uneven distribution of rain over the country each year, water resources available for utilization, including ground water, is claimed to be about 1122 billion cubic meters. Much of this water is unsafe, because pollution degrades water quality.

Water pollution severely limits the amount of water available to Indian consumer, its industry and its agriculture.

## THE GANGES

To know why 1,000 Indian children die of diarrhoeal sickness every day, take a wary stroll along the Ganges in Varanasi. As it enters the city, Hinduism's sacred river contains 60,000 faecal coliform bacteria per 100 millilitres, 120 times more than is considered safe for bathing. Four miles downstream, with inputs from 24 gushing sewers and 60,000 pilgrim-bathers, the concentration is 3,000 times over the safety limit. In places, the Ganges becomes black and septic. Corpses, of semi-cremated adults or enshrouded babies, drift slowly by. — *The Economist on December 11, 2008*

More than 400 million people live along the Ganges River. An estimated 2,000,000 persons ritually bathe daily in the river, which is considered holy by Hindus. Ganges river pollution from sewage and semi-cremated remains is thus a major health risk.

## THE YAMUNA

NewsWeek describes Delhi's sacred Yamuna River as "a putrid ribbon of black sludge" where fecal bacteria is 10,000 over safety limits despite a 15-year programme to address the problem. Cholera epidemics are not unknown. NewsWeek observes India's messy democracy is particularly ill equipped to handle the conflicting pressures of rapid growth and poverty. Even though India revised its national water policy in 2002 to encourage community participation and decentralize water management, the country's Byzantine bureaucracy ensures that it remains a "mere statement of intent." Responsibility for managing water issues is fragmented among a dozen different ministries and departments without any coordination. The government bureaucracy and state-run project department has failed to solve the problem, despite having spent many years and $500 million on this project.

# METHODS OF WATER POLLUTION

Methods of controlling water pollution fall into three general categories: physical, chemical, and biological. For example, one form of water pollution consists of suspended solids such as fine dirt and dead organisms. These materials can be removed from water by simply allowing the water to sit quietly for a period of time, thereby allowing the pollutants to settle out, or by passing the water through a filter. (The solid pollutants are then trapped in the filter.) Chemical reactions can be used to remove pollutants from water. For example, the addition of alum (potassium aluminum sulfate) and lime (calcium hydroxide) to water results in the formation of a thick, sticky precipitate. When the precipitate begins to settle out, it traps and carries with it solid particles, dead bacteria, and other components of polluted water. Biological agents can also be used to remove pollutants from water. Aerobic bacteria (those that need oxygen to survive) and anaerobic bacteria (those that do not require oxygen) attack certain chemicals in polluted water and convert them to a harmless form.

## SOLID POLLUTANTS

Solid pollutants consist of garbage, sewage sludge, paper, plastics, and many other forms of waste materials. One method of dealing with solid pollutants is simply to bury them in dumps or landfills. Another approach is to compost them, a process in which microorganisms turn certain types of pollutants into useful fertilizers. Finally, solid pollutants can also be incinerated (burned).

## TRADE AND ENVIRONMENT

The main goal is to make the complex relationship between the environment and international trade more understandable from the developing countries' perspective. It also aims to dispel the idea that the relationship between trade and environment can easily be described as either negative or positive.

It is extremely complex and varies from country to country. There are both threats and opportunities in this relationship for developing countries pursuing economic development and environmental protection. The conclusions that can be drawn from this that from the developing countries' perspective is to exploit the opportunities and reduce the threats, and in so doing, to maximize the net positive contribution that trade can make sustainable development. A broader and clearer understanding of the linkages between trade, environment, and development is a prerequisite for seizing those opportunities and reducing those threats.

On the other hand, environmental requirements however, may be used for protectionist purposes and may operate as a non-tariff barrier. In future, there is a likelihood of more trade embargos in the name of environment, especially in cases where countries fail to comply with certain environmental standard requirements. Thus, the outcome of the various case studies has the potential for strengthening the hand of protectionist lobbies, in the name of conservation and protection of the environment.

Finally, the threat of embargo can also be used to make countries adhere to certain environmental policies and regulations, which they would otherwise not adopt due to various socio-economic compulsions. Moreover, these environmental friendly policies demand the attention and cooperation of international institutions such as the World Bank as well as NGOs and other grassroots environmental groups. Protecting the environment is a luxury good, often only afforded to the wealthy countries.

Therefore, these policies should be coupled with ones that encourage the transfer and exchange of environmentally friendly technologies and the build up of industries oriented to sustainable development.

## WATER QUALITY CONSERVATION

India is endowed with diverse geological formations from oldest achaeans to recent alluviums and characterized by varying climatic conditions in different parts of the country. The natural chemical content of groundwater is influenced by depth of the soils and sub-surface geological formations through which groundwater remains in contact.

In general, greater part of the country, groundwater is of good quality and suitable for drinking, agricultural or industrial purposes. Groundwater in shallow aquifers is generally suitable for use for different purposes and is mainly of Calcium Bicarbonate and mixed type. However, other types of water are also available including Sodium-Chloride water. The quality in deeper aquifers also varies from place to place and is generally found suitable for common uses. There is salinity problem in the coastal tracts and high incidence of fluoride, arsenic, iron and heavy metals etc. in isolated pockets have also been reported. The main groundwater quality problems in India are as follows.

### SALINITY

Salinity in groundwater can be of broadly categorized into two types, *i.e.* inland salinity and coastal salinity.

### INLAND SALINITY

Inland salinity in groundwater is prevalent mainly in the arid and semi-arid regions of Rajasthan, Haryana, Punjab and Gujarat, Uttar Pradesh, Delhi, Andhra Pradesh, Maharashtra, Karnataka and Tamil Nadu.

There are several places in Rajasthan and southern Haryana where EC values of groundwater is quite high making water non-potable. In some areas of Rajasthan and Gujarat, groundwater salinity is so high that the well water is directly used for salt manufacturing by solar evaporation. Inland salinity is also caused due to practice of surface water irrigation without consideration of groundwater status. The gradual rise of groundwater levels with time has resulted in water logging and heavy evaporation in semi-arid regions lead to salinity problem in command areas.

### COASTAL SALINITY

The Indian subcontinent has a dynamic coast line of about 7500 km length. It stretches from Rann of Kutch in Gujarat to Konkan and Malabar coast to Kanyakumari in the south to northwards along the Coromandal coast to Sunderbans in West Bengal.

The western coast is characterized by wide continental shelf and is marked by backwaters and mud flats while the eastern coast has a narrow continental shelf and is characterized by deltaic and estuarine land forms. Groundwater in coastal areas occurs under unconfined to confined conditions in a wide range of unconsolidated and consolidated formations. Normally, saline water bodies owe their origin to entrapped sea water (connate water), sea water ingress, leachates from navigation canals constructed along the coast, leachates from salt pans etc. In India, salinity problems have been observed in a number of places in most of the coastal states of the country. Problem of salinity ingress has been conspicuously noticed in Minjur area of Tamil Nadu and Mangrol – Chorwad-Porbander belt along the Saurashtra coast. Under Hydrology Project (PhaseII), National Institute of Hydrology, Roorkee has taken-up a purpose driven study related to seawater intrusion in Minsar river basin in Porbandar district of Gujarat.

**FLUORIDE**

85 per cent of rural population of the country uses groundwater for drinking and domestic purposes. High concentration of fluoride in groundwater beyond the permissible limit of 1.5 mg/l poses the health problem. The occurrences of fluoride beyond permissible limit (> 1.5 mg/l) has been observed based on the chemical analysis of water samples collected from the groundwater observation wells.

**ARSENIC**

The occurrence of arsenic in groundwater was first reported in 1980 in West Bengal. In West Bengal, 79 blocks in 8 districts have arsenic beyond the permissible limit of 0.05 mg/l. The most affected districts are on the eastern side of Bhagirathi river in the districts of Malda, Murshidabad, Nadia, North 24 Parganas and South 24 Parganas and western side of the districts of Howrah, Hugli and Bardhman. The occurrence of arsenic in groundwater is mainly in the intermediate aquifers upto the depth of 100 m. The deeper aquifers are free from arsenic contamination. Apart from West Bengal, arsenic contamination in groundwater has been found in the states of Bihar, Chhattisgarh, Uttar Pradesh and Assam. Arsenic in groundwater has been reported in 15 districts In Bihar, 9 districts in U.P. and one district each in Chhatisgarh and Assam states. The occurrence of arsenic in the states of Bihar, West Bengal and Uttar Pradesh is in alluvium formation but in the state of Chhattisgarh, it is in the volcanic rocks exclusively confined to N-S trending Dongargarh-Kotri ancient rift zone. It has also been reported in Dhemaji district of Assam.

**IRON**

High concentration of iron (>1.0 mg/l) in groundwater has been observed in more than 1.1 lakh habitations in the country. Groundwater contaminated by iron has been reported from the states of Andhra Pradesh, Assam, Bihar, Chhattisgarh, Goa, Gujarat, Haryana, J&K, Jharkhand, Karnataka, Kerala, Madhya Pradesh, Maharashtra, Manipur, Meghalaya, Orissa, Punjab, Rajasthan, Tamil Nadu, Tripura, Uttar Pradesh, West Bengal and UT of Andaman & Nicobar.

### NITRATE

Nitrate is a very common constituent in the groundwater, especially in shallow aquifers. The source is mainly from anthropogenic activities. High concentration of nitrate in water beyond the permissible limit of 45 mg/l causes health problems. High nitrate concentration in groundwater in India has been found in almost all hydrogeological formations.

Implementation of water pollution prevention strategies and restoration of ecological systems are integral components of all development plans. To preserve our water and environment, we need to make systematic changes in the way we grow our food, manufacture the goods, and dispose off the waste. In India, agriculture is the biggest user and polluter of water. If pollution by agriculture is reduced, it would improve water quality and would also eliminate cost incurred for treatment of diseases. Like all other inputs, there is an optimal quantity of fertilizer for given conditions and excess application does not improve the crop yield. Pricing of fertilizers and pesticides as well as appropriate legislation to regulate their use will also go a long way in stopping indiscriminate use. Industries need to carefully treat their waste discharges. Manufacturers may reduce water pollution by reusing materials and chemicals and switching over to less toxic alternatives. Industrial symbiosis, in which the unusable wastes from one product/firm become the input for another, is an attractive solution. Also, there is a need to encourage reductions or replacement of toxic chemicals, possibly through fiscal measures.

## URBAN WATER SUPPLY AND SANITATION

The urban water supply and sanitation sector in the country is suffering from inadequate levels of service, an increasing demand-supply gap, poor sanitary conditions and deteriorating financial and technical performance. Supply of water is highly erratic and unreliable.

Transmission and distribution networks are old and poorly maintained, and generally of a poor quality. Consequently physical losses are typically high, ranging from 25 to over 50 per cent. Low pressures and intermittent supplies allow back siphoning, which results in contamination of water in the distribution network. Water is typically available for only 2-8 hours a day in most Indian cities. The situation is even worse in summer when water is available only for a few minutes, sometimes not at all.

Sanitation and water management should be looked at simultaneously. Too often attention is focused on drinking water supply, leaving sanitation and wastewater treatment for later. However, for every 100 litres of water going into a house about 90 litres will have to leave the plot again. Water supply is an institutional process and an institutional framework for effective water supply and sanitation has to comply with the functions of policy, regulation and sector organization, management of quality, infrastructure and on-site sanitation.

It is desirable to think of water supply, sanitation and wastewater in an integrated way. Urban centres in India are facing an ironical situation today. On one hand, there is the acute water scarcity and on the other, the streets are often flooded during the monsoons. This has led to serious problems with quality and quantity of groundwater. This is despite the fact that all these cities receive good rainfall.

However, this rainfall occurs during short spells of high intensity. Because of such short duration of heavy rain, most of the rain falling on the surface tends to flow away rapidly leaving very little for recharge of groundwater. As water shortage increases, alternative sources of water supply are gaining importance. These include sewage recycle, rainwater harvesting, etc. It should be made mandatory for each industry to install water management solutions to recycle its waste water for reuse. Major step in this front is through the development of industrial effluent recycle solution which integrates physio-chemical, biological and membrane separation processes for optimum water recovery. They achieve water management through water recycle and source reduction, and waste management through product recovery and waste minimization.

## PEOPLE PARTICIPATION AND CAPACITY BUILDING

For making the people of various sections of the society aware about the different issues of water resources management, a participatory approach may be adopted. Mass communication programmes may be launched using the modern communication means for educating the people about water conservation and efficient utilization of water. Capacity building should be perceived as the process whereby a community equips itself to become an active and well-informed partner in decision making. The process of capacity building must be aimed at both increasing access to water resources and changing the power relationships between the stakeholders. Capacity building is not only limited to officials and technicians but must also include the general awareness of the local population regarding their responsibilities in sustainable management of the water resources. Policy decisions in any water resources project should be directed to improve knowledge, attitude and practices about the linkages between health and hygiene, provide higher water supply service levels and to improve environment through safe disposal of human waste.

Sustainable management of water requires decentralized decisions by giving authority, responsibility and financial support to communities to manage their natural resources and thereby protect the environment.

## GROUNDWATER POLLUTION

Interactions between groundwater and surface water are complex. Consequently, groundwater pollution, sometimes referred to as groundwater contamination, is not as easily classified as surface water pollution. By its very nature, groundwater aquifers are susceptible to contamination from sources that may not directly affect surface water bodies, and the distinction of point vs. non-point source may be irrelevant.

A spill of a chemical contaminant on soil, located away from a surface water body, may not necessarily create point source or non-point source pollution, but nonetheless may contaminate the aquifer below. Analysis of groundwater contamination may focus on soil characteristics and hydrology, as well as the nature of the contaminant itself.

## CAUSES OF WATER POLLUTION

When you consider the many different causes of water pollution, it's not difficult to understand why drinking water contamination is such a widespread problem. Water pollution occurs both when harmful substances directly enter the water supply, as well as through changes that occur in the environment as a result of pollution causing activities. When you realize how many different types of water pollution there are, it's easy to see why water filtration systems are a necessity for every home.

## COMMON CAUSES OF WATER POLLUTION

### OIL

Petroleum often pollutes water in the form of oil. Oil spills from ships and super- tankers, and from off-shore oil drilling operations cause pollution. Oil and petrol that leaks from cars and trucks also washes off roads and into waterways through storm water drains. Oil forms a thin layer on top of water and act like a lid on the surface and the water. Animals and plants living in the water can't breathe, the oil coats the feathers of water birds, and the fur of animals that swim in the water, causing them to become sick and, if there is a great amount of oil on their bodies, to die. Even the insects that live on the surface of the water are badly affected.

### FERTILIZERS

Fertilizers contain nutrients such as nitrates and phosphates that help plants to grow. That's why farmers use them. When fertilizers are washed into rivers and streams the nitrates and phosphates cause excessive growth of water plants. The plants clogs the waterways, use up oxygen in the water, and block light to deeper waters. This is harmful to the fish and other invertebrates that live in water because it make it hard for the animals to breathe.

## SOIL

Pollution of waterways is also caused when silt and soil washes off ploughed fields, construction and logging sites, and from river banks when it rains.

## SEWAGE AND OTHER ORGANIC POLLUTANTS

When material such as leaves and grass clippings, and waste from farm animals enters the water, it rots and breaks down and uses up the oxygen in the water. Many types of fish and other aquatic animals cannot survive. Organisms such as bacteria and viruses enter waterways through untreated sewage in storm-water drains, run-off from septic tanks, and from boats whose owners dump sewage into the water. These microscopic pollutants cause sickness in people and in animals that drink or live in the water.

## CHEMICALS

Chemical pollution entering rivers and streams causes great destruction. The chemicals can come from factories, construction sites, mining operations, and from homes when people pour chemicals down the sink or down the toilet.

## PLASTICS

Floating plastic is ugly, and harmful to the environment. Plastic rubbish is not biodegradable (it doesn't rot away after we have used it) It can choke animals that try to eat it, and drown those that get tangled in it.

## LITTER

When people drop litter such as plastic and cans, food wrappers and cigarette butts, they can be washed by the rain into rivers and other waterways through storm water drains in the streets. At the beach, it is important that people take home their litter or put it into garbage bins at the beach so that it doesn't get into the sea.

### Other Causes of Water Pollution Include:

- Air Pollution
- Carbon Dioxide
- Fuel Additives
- Increased Water Temperature
- Industrial Development
- Landfills
- Mining Activities

- Gasoline
- Household Cleaning Products
- Personal Care Products
- Pesticides
- Pharmaceuticals
- Sediment.

## THE EFFECTS OF WATER POLLUTION

The effects of water pollution are far-reaching. And it's not only humans who are affected. All plants and animals must endure poisonous drinking water, river and lake ecosystems that have become unbalanced and can no longer support biodiversity.

Deforestation from acid rain can occur as well. On the whole, water pollution has long-term effects on our health and economic productivity. Fight water pollution to keep our planet safe. The effects of water pollution differ from region to region, depending on the pollutants in the water and environmental factors. Common effects of water pollution include unhealthy or poisonous water, sick animals that pass their sickness on to humans, ecosystems that are unable to support a normal diverse animal and plant habitat and more.

### POLLUTION AFFECTS THE FOOD CHAIN

Pollution in the form of organic material enters waterways in many different forms as sewage, as leaves and grass clippings, or as run-off from livestock feedlots and pastures. When natural bacteria and protozoan in the water break down this organic material, they destroyed the whole things.

### POLLUTION AFFECT AQUATIC ECOSYSTEMS

In nature nothing exists alone. Living things relate to each other as well as to their non-living, but supporting, environments.These complex relationships are called ecosystems. Each body of water is a delicately balanced ecosystem in continuous interaction with the surrounding air and land. Whatever occurs on the land and in the air also affects the water. If a substance enters a river or lake, the water can purify itself biologically — but only to a degree. Whether it is in the smallest stream or lake — or even in the mighty oceans — the water can absorb only so much. It reaches a point where the natural cleaning processes can no longer cope. For example, various species of fish now suffer from tumors and lesions, and their reproductive capacities are decreasing

### POLLUTION AFFECT MARINE LIFE

Lakes many are persistent toxic chemicals such as DDT. Populations of fish consuming birds and mammals also seem to be on the decline. Of the ten most highly valued species of fish in Lake Ontario, seven have now almost totally vanished. People dump hazardous materials into the ocean to get rid of them. Sewage and waste from factories and cities can reach the ocean. This pollution is very harmful. It can kill the plants and animals living in the ocean. Dangerous chemicals like mercury kill the aquatic environment very quickly.

## WATER POLLUTION'S CHAIN REACTION

One of the many causes of water pollution is sewage and fertilizers that contain nutrients such as nitrates and phosphates. When these enter our water system in excess levels, the growth of aquatic plants and algae is over-stimulated. As a result of the excessive growth of these aquatic plants, our waterways are clogged. They use up dissolved oxygen as they decompose, and block light to deeper waters. Subsequently, the respiration ability or fish and other invertebrates that reside in water are also damaged.

## HEALTH IMPACTS OF WATER POLLUTION

It is a well-known fact that clean water is absolutely essential for healthy living. Adequate supply of fresh and clean drinking water is a basic need for all human beings on the earth, yet it has been observed that millions of people worldwide are deprived of this. Freshwater resources all over the world are threatened not only by over exploitation and poor management but also by ecological degradation. The main source of freshwater pollution can be attributed to discharge of untreated waste, dumping of industrial effluent, and run-off from agricultural fields. Industrial growth, urbanization and the increasing use of synthetic organic substances have serious and adverse impacts on freshwater bodies. It is a generally accepted fact that the developed countries suffer from problems of chemical discharge into the water sources mainly groundwater, while developing countries face problems of agricultural run-off in water sources. Polluted water like chemicals in drinking water causes problem to health and leads to water-borne diseases which can be prevented by taking measures can be taken even at the household level.

## GROUNDWATER AND ITS CONTAMINATION

Many areas of groundwater and surface water are now contaminated with heavy metals, POPs (persistent organic pollutants), and nutrients that have an adverse affect on health. Water-borne diseases and water-caused health problems are mostly due to inadequate and incompetent management of water resources. Safe water for all can only be assured when access, sustainability, and equity can be guaranteed. Access can be defined as the number of people who are guaranteed safe drinking water and sufficient quantities of it. There has to be an effort to sustain it, and there has to be a fair and equal distribution of water to all segments of the society. Urban areas generally have a higher coverage of safe water than the rural areas. Even within an area there is variation: areas that can pay for the services have access to safe water whereas areas that cannot pay for the services have to make do with water from hand pumps and other sources.

In the urban areas water gets contaminated in many different ways, some of the most common reasons being leaky water pipe joints in areas where the water pipe and sewage line pass close together. Sometimes the water gets polluted at source due to various reasons and mainly due to inflow of sewage into the source.

*Pesticides*. Run-off from farms, backyards, and golf courses contain pesticides such as DDT that in turn contaminate the water. Leechate from landfill sites is another major contaminating source. Its effects on the ecosystems and health are endocrine and reproductive damage in wildlife. Groundwater is susceptible to contamination, as pesticides are mobile in the soil. It is a matter of concern as these chemicals are persistent in the soil and water.

*Sewage*. Untreated or inadequately treated municipal sewage is a major source of groundwater and surface water pollution in the developing countries. The organic material that is discharged with municipal waste into the watercourses uses substantial oxygen for biological degradation thereby upsetting the ecological balance of rivers and lakes. Sewage also carries microbial pathogens that are the cause of the spread of disease.

*Nutrients*. Domestic waste water, agricultural run-off, and industrial effluents contain phosphorus and nitrogen, fertilizer run-off, manure from livestock operations, which increase the level of nutrients in water bodies and can cause eutrophication in the lakes and rivers and continue on to the coastal areas. The nitrates come mainly from the fertilizer that is added to the fields. Excessive use of fertilizers cause nitrate contamination of groundwater, with the result that nitrate levels in drinking water is far above the safety levels recommended. Good agricultural practices can help in reducing the amount of nitrates in the soil and thereby lower its content in the water.

*Synthetic organics*. Many of the 100 000 synthetic compounds in use today are found in the aquatic environment and accumulate in the food chain. POPs or Persistent organic pollutants, represent the most harmful element for the ecosystem and for human health, for example, industrial chemicals and agricultural pesticides. These chemicals can accumulate in fish and cause serious damage to human health. Where pesticides are used on a large-scale, groundwater gets contaminated and this leads to the chemical contamination of drinking water.

*Acidification*. Acidification of surface water, mainly lakes and reservoirs, is one of the major environmental impacts of transport over long distance of air pollutants such as sulphur dioxide from power plants, other heavy industry such as steel plants, and motor vehicles. This problem is more severe in the US and in parts of Europe.

## CHEMICALS IN DRINKING WATER

Chemicals in water can be both naturally occurring or introduced by human interference and can have serious health effects.

*Fluoride*. Fluoride in the water is essential for protection against dental caries and weakening of the bones, but higher levels can have an adverse effect on health. In India, high fluoride content is found naturally in the waters in Rajasthan.

*Arsenic*. Arsenic occurs naturally or is possibly aggrevated by over powering aquifers and by phosphorus from fertilizers. High concentrations of arsenic in water can have an adverse effect on health.A few years back, high concentrations of this element was found in drinking water in six districts in West Bengal. A majority of people in the area was found suffering from arsenic skin lesions. It was felt that arsenic contamination in the groundwater was due to natural causes. The government is trying to provide an alternative drinking water source and a method through which the arsenic content from water can be removed.

*Lead*. Pipes, fittings, solder, and the service connections of some household plumbing systems contain lead that contaminates the drinking water source.

*Recreational use of water*. Untreated sewage, industrial effluents, and agricultural waste are often discharged into the water bodies such as the lakes, coastal areas and rivers endangering their use for recreational purposes such as swimming and canoeing.

*Petrochemicals*. Petrochemicals contaminate the groundwater from underground petroleum storage tanks.

*Other heavy metals*. These contaminants come from mining waste and tailings, landfills, or hazardous waste dumps.

*Chlorinated solvents*. Metal and plastic effluents, fabric cleaning, electronic and aircraft manufacturing are often discharged and contaminate groundwater.

## DISEASES

Water-borne diseases are infectious diseases spread primarily through contaminated water. Though these diseases are spread either directly or through flies or filth, water is the chief medium for spread of these diseases and hence they are termed as water-borne diseases.

Most intestinal (enteric) diseases are infectious and are transmitted through faecal waste. Pathogens – which include virus, bacteria, protozoa, and parasitic worms – are disease-producing agents found in the faeces of infected persons. These diseases are more prevalent in areas with poor sanitary conditions. These pathogens travel through water sources and interfuses directly through persons handling food and water. Since these diseases are highly infectious, extreme care and hygiene should be maintained by people looking after an infected patient. Hepatitis, cholera, dysentery, and typhoid are the more common water-borne diseases that affect large populations in the tropical regions.

A large number of chemicals that either exist naturally in the land or are added due to human activity dissolve in the water, thereby contaminating it and leading to various diseases.

*Pesticides.* The organophosphates and the carbonates present in pesticides affect and damage the nervous system and can cause cancer. Some of the pesticides contain carcinogens that exceed recommended levels. They contain chlorides that cause reproductive and endocrinal damage.

*Lead.* Lead is hazardous to health as it accumulates in the body and affects the central nervous system. Children and pregnant women are most at risk.

*Fluoride.* Excess fluorides can cause yellowing of the teeth and damage to the spinal cord and other crippling diseases.

*Nitrates.* Drinking water that gets contaminated with nitrates can prove fatal especially to infants that drink formula milk as it restricts the amount of oxygen that reaches the brain causing the 'blue baby' syndrome. It is also linked to digestive tract cancers. It causes algae to bloom resulting in eutrophication in surface water.

*Petrochemicals.* Benzene and other petrochemicals can cause cancer even at low exposure levels.

*Chlorinated solvents.* These are linked to reproduction disorders and to some cancers.

*Arsenic.* Arsenic poisoning through water can cause liver and nervous system damage, vascular diseases and also skin cancer.

Other heavy metals. –Heavy metals cause damage to the nervous system and the kidney, and other metabolic disruptions.

*Salts.* It makes the fresh water unusable for drinking and irrigation purposes.

Exposure to polluted water can cause diarrhoea, skin irritation, respiratory problems, and other diseases, depending on the pollutant that is in the water body. Stagnant water and other untreated water provide a habitat for the mosquito and a host of other parasites and insects that cause a large number of diseases especially in the tropical regions. Among these, malaria is undoubtedly the most widely distributed and causes most damage to human health.

## PREVENTIVE MEASURES

Water-borne epidemics and health hazards in the aquatic environment are mainly due to improper management of water resources. Proper management of water resources has become the need of the hour as this would ultimately lead to a cleaner and healthier environment.

In order to prevent the spread of water-borne infectious diseases, people should take adequate precautions. The city water supply should be properly checked and necessary steps taken to disinfect it. Water pipes should be regularly checked for leaks and cracks. At home, the water should be boiled, filtered, or other methods and necessary steps taken to ensure that it is free from infection.

## MINAMATA: ENVIRONMENTAL CONTAMINATION WITH METHYL MERCURY

In Minamata, Japan, inorganic mercury was used in the industrial production of acetaldehyde. It was discharged into the nearby bay as waste water and was ingested by organisms in the bottom sediments. Fish and other creatures in the sea were soon contaminated and eventually residents of this area who consumed the fish suffered from MeHg (methyl mercury) intoxication, later known as the Minamata disease. The disease was first detected in 1956 but the mercury emissions continued until 1968. But even after the emission of mercury stopped, the bottom sediment of the polluted water contained high levels of this mercury.

Various measures were taken to deal with this disease. Environmental pollution control, which included cessation of the mercury process; industrial effluent control, environmental restoration of the bay; and restrictions on the intake of fish from the bay. This apart research and investigative activities were promoted assiduously, and compensation and help was offered by the Japanese Government to all those affected by the disease.

The Minamata disease proved a turning point, towards progress in environment protection measures. This experience clearly showed that health and environment considerations must be integrated into the process of economic and industrial development from an early stage.

## POLLUTION OF DRINKING WATER

Water is an important factor in the life of organisms. It supports life system and its shortage has been to main concerns of human beings.

Environmentalists and social scientists are warning about the availability of fresh water to human consumption after 2020 AD. Almost all the developing countries are going to face shortage of fresh water for human consumption by 2020 AD. Industrialization, development of civilization and population growth have increased dependence on natural resources, which are fast depleting. Availability of fresh water is a must for human consumption. The fresh water resources are fast depleting or either getting polluted due to careless handling of water resources by the developing countries. The present paper mainly deals with the fresh water pollution in irrigation canals of Godavari delta System from its source of generation (Dowalaiswaram Barrage) upto the field input.

Generally the canal water in Godavari delta system will be utilized for water requirements of irrigable crops as well as for drinking purpose in the region. As regards pollution of fresh water in irrigation canals of Godavari Delta System, two important aspects are to be considered. One is pollution of canal water from source of generation (Dowalaiswaram barrage) to field input, while the fresh water is being carried through irrigation canals. Another is due to pollution caused to canal water by disorderly development of prawn/fish culture in the region. Apart from the above, there are industrial pollutants (due to large scale industrialization) etc. while they are being allowed to flow into irrigation canals and drains causes pollution.

## POLLUTION OF CANAL WATER

Generally the canal water, from reservoirs or barrages are drawn through main canal, and distributed through branch canals, distributaries, field channels to the field. As far as Godavari delta is concerned, the irrigation canals and branch canals are passing through many of the towns, villages, hamlets, hutments etc. The main source of drinking water in the villages is through village drinking water tanks, fed by irrigation water.

Civilization and Industrialization has made mans pursuit for development more difficult. From the time immemorial, the drinking water tanks are source of fresh water for drinking purpose in the villages, in preference to urban areas. Now the position has changed. They are no more fresh water ponds, but containing polluted canal water only. The water pollution even in the villages is predominantly high; some times they also have to depend on ground water or private water resources for drinking purposes. This is due to large scale water pollution in canals from its source of generation to field input.

All the irrigation canals are passing by the towns, villages, hamlets hutments, carrying fresh water. There is considerable waste land *i.e.* government or poramboke (canal poramboke or drain poramboke) available on either side of delta canals or drains, allowing for possible widening of canals/ drains at a later stage. Thanks to the tbresightedness of "Apara Bhagiratha" Sir Arthur Cotton, the Eminent British Engineer, who constructed Godavari Anicut (old) and canal system in the year 1854.

At present all the canal banks and berms are encroached and occupied by local people for construction of dwelling houses, pucca lavatories etc. thus allowing all human waste/effluents into fresh water canal. The canal banks and berths are the open lavatories on the village nearby. For many of the villages and towns, the main outlets of drainage and sewage water are irrigation canals only.

The irrigation canals which are passing nearby the villages and towns are the dumping yards for the human and animal wastes sewage etc. In some places, the medical and surgical wastes and industrial wastes are dumped into fresh water canals. By the time the water reaches the tail end, it is contaminated with the full of the said effluents.

The million dollar questions asked now are: 1) Who is responsible for the above situation? 2. What are the existing Acts/laws to control the situation? 3. What are the remedial measures? The people at large and encroachers and hutment dwellers who have encroached into govt. land/poramboke in particular are responsible for the above situation.

The water (Prevention and control of pollution) Act, 1974 authorized central government to constitute The central pollution control Board" to exercise the powers conferred on and to perform the functions assigned to the board under the Act, authorized the State government, to constitute State Boards. If the canal water are to be made, free from pollution, the encroachments on the either side of the canals and canal berms (public lavatories) are to be evicted. All the sewage water freely flowing into the canal is to be stopped. Public awareness is to be created among the people not to throw objectionable matter such as garbage, sewage waste, human and animal wastes etc. into the irrigation canals. It is a socio-economic problem; the State government must take } effective measures.

Supreme Court in M.C. Mehta Vs. Union of India in Olium Gas Leakage case (AIR 1987, SC 1086) have evolved 'the Rule of absolute Liability" developed from the rule of strict liability and awarded damages to the victims. The wrong doers arc to be punished under 'absolute liability' for which no defence is available. The same principle must be applied in case of offenders of water pollution, who ever may be.

## POLLUTION OF FRESH WATER

Seafood has created a sensation in the world trade because of its health attitudes. It is the fastest moving commodity with persistent demand and high unit value. Seafood sector is growing into a multi billion dollar industry, with immense potential for the future. Prawns are the largest contribution to Indian fishing exports, accounting for an export share of 22 in volume, 48 in value. Prawns exports have been growing steadily from India in the recent years, mainly due to increase in production from capture as well as from culture. The export of marine products from India amounted to 3500 crores in the year 1995-96 and it stood at 6400 crores in the year 2000-01. Sea food exports alone constitute about 3.14 of the gross export earnings of our country, which is a vital to the Indian economy.

In view of the good return on prawn/fish culture, many of the irrigation lands under paddy cultivation are being converted into prawn/fish tanks, in Godavari Delta, including that of "Coastal Regulation Zone". This indiscriminate conversion of paddy fields is causing water pollution, including imbalance in eco-system.

The question of prohibiting or regulating aquaculture, prawn culture and shrimp culture came up for consideration for the Apex Court in S. Jagannath Vs. Union of India (AIR 1997 SC 811).

The court gave the following directions in this case:

1. We declare that the notification issued in G.O.Ms.No.120, dated 4.10.1999 is valid.
2. The respondents shall forthwith take adequate steps for stoppage and regulation of effluents discharged from the industries and municipalities into Kolleru Lake and strictly adhere to the standards laid down by the Ministry of Environment, Government of India for the purpose of preservation and maintenance of the lake and ecology in accordance with law and state shall make all endeavours to bring back Kolleru Lake to its pristine glory.
3. No Pisci culture/aquaculture/shrimp culture should be permitted to be undertaken with the Kolleru lake sanctuary and only traditional methods of fishing as directed in G.o.ms.No.120, dated 4.10.1999 should be permitted. Any person intending to take recourse to aqua culture or Pisci culture or shrimp culture must file requisite applications before the appropriate aqua culture authority provided their lands fell outside the notification-dated 4.10.1999 and the area of sanctuary.
4. State shall ensure for removal of all encroachments of Kolleru Lake bed area in accordance with G.O.ms.No.l20, dt.4.10.1999. However, the enforcement authorities failed to remove encroachments in Kolleru Lake, further the authorities tailed to check conversion of agriculture lands into aqua culture, though such activities are going on without obtaining permissions from the proper authorities.

The method of irrigation adopted in Godavari Delta is by 'flooding method' mainly of 'strip border method". In this the irrigable land is levelled and divided into suitable size along plots by 0.30 mt. high field lands into a number of long and narrow strips. Water is allowed at the head or upper end of each strip and it flows along the strip in the form of a thin 5 to 7.5 cm stretch of water to the lower end of the strip. The irrigation system is mainly from field to field initially. The drainage water from one plot is collected and discharged into field drainage. The drainage system is not pucca as that of canal system. With the result the drainage water from upper field will flood over the adjacent field to join to nearest drainage course. Under this process of field to field irrigation, the drainage water of upper field will be input for lower field. This process of water utilization will not cause any effect so far as all the field block is under cultivation. If some land is under prawn/fish culture, the drainage water of prawn/fish tanks will be input for irrigation land. The problem has been aggravated due to smaller plots of land holding by the ryots, in Godavari delta.

Due to successive use of same water for different tanks, apart from change of physical and chemical properties. The water will acquire 'dielectric characteristics'. Ultimately the water gets polluted and will not be suitable for drinking purpose.

How the questions that arise are:

1. How far it is admissible to allow prawn/fish culture to interfere with the irrigation system developed exclusively for crop water requirements?
2. Can government centre and state, ban prawn/fish culture which are hazardously polluting water for human consumption in view of the tremendous foreign exchange earning capacity?

It is a socio-economic problem. No Government can ill afford to loose the tremendous foreign exchange earnings out of sea food exports. More over small and marginal farmers are dependent on culture. The only solution seems to be segregation of total irrigable land under delta system into two zones, one for raising irrigable crops and another for cultivating prawn/fish tanks.

The lands or areas surrounded by fisherman villages must be classified as zones for culture etc. The remodelling of the Irrigation system can be taken up by "Marine products Export Development Authority out of the funds available with them, by way of Export Cess. The irrigation canal water is polluted, before it reaches the consumers in the villages for drinking purposes. Art. 21 of the constitution which guarantees Right to life. The Constitutional Review Committee while suggesting certain additions to Art. 21 (*i.e.* 21A to 21C) ignored to include "right to have a clean environment" under Art 21 therefore this may be incorporated as Art. 21 D. Thus making it as a fundamental Right.

Both central and state government and people in particular are responsible for the large scale pollution of canal water. It is a violation of fundamental rights, guaranteed to the citizen. Then the judiciary must act upon, when the executive is inactive. It is obvious that "the Water (Prevention and Control of Pollution Act 1974" even though enacted as far back as 1974 could not yield concrete results, so far obviously.

The supreme court has banned raising of culture in 'Coastal Regulation Zone". If the constitutional rights of getting free water and air to be enforced to the citizens, still more stringent laws to be enacted. A statue which will give an immediate action will only serve the purpose. The Water Act, 1974 and Environment (Protection) Act, 1986 may be amended which empower the pollution control Boards to take appropriate action instead of filing the cases in courts. Further it should also enable a magistrate atleast First Class Magistrate to initiate action 'suo moto' without waiting for Board to initiate action.

## RECENT EFFORTS TO ADDRESS WATER ISSUES

The government of India envisions a US$154 billion project to interlink all major river networks in India. This initiative would connect water-deficient areas to water-abundant ones by interlinking 37 Indian rivers. One of the largest projects anywhere in the world, it would transfer water through 30 links across 9,600 kilometres. It would connect 32 dams and use 56 million tons of cement and 2 million tons of steel. The project aims a transformation of India's water treatment, management, transmission and distribution. The Indian government has proposed reforms to attract investment and privatization of its water networks. Water companies from all over the world have established a presence in India to pursue an estimated 70 projects worth several billion dollars in 20 Indian cities. India is debating the social and environmental impact of this project. One of the first projects under consideration is the linking of Ken and Betwa rivers in northern India.

## AIR POLLUTION

### FUEL WOOD AND BIOMASS BURNING

Fuelwood and biomass burning is the primary reason for near-permanent haze and smokeobserved above rural and urban India, and in satellite pictures of the country. Fuelwood and biomass cakes are used for cooking and general heating needs. These are burnt in cook stoves known as *chullah* or *chulha* in some parts of India. These cook stoves are present in over 100 million Indian households, and are used two to three times a day, daily. As of 2009, majority of Indians still use traditional fuels such as dried cow dung, agricultural wastes, and firewood as cooking fuel. This form of fuel is inefficient source of energy, its burning releases high levels of smoke, PM10 particulate matter, NOX, SOX, PAHs, polyaromatics, formaldehyde, carbon monoxide and other air pollutants. Some reports, including one by the World Health Organization, claim 300,000 to 400,000 people die of indoor air pollution and carbon monoxide poisoning in India because of biomass burning and use of chullahs. Burning of biomass and firewood will not stop, unless electricity or clean burning fuel and combustion technologies become reliably available and widely adopted in rural and urban India.

India is the world's largest consumer of fuelwood, agricultural waste and biomass for energy purposes. From the most recent available nationwide study, India used 148.7 million tonnes coal replacement worth of fuelwood and biomass annually for domestic energy use.

India's national average annual per capita consumption of fuel wood, agri wate and biomass cakes was 206 kilogram coal equivalent. In 2010 terms, with India's population increased to about 1.2 billion, the country burns over 200 million tonnes of coal replacement worth of fuel wood and biomass every year to meet its energy need for cooking and other domestic use. The study found that the households consumed around 95 million tonnes of fuelwood, one-third of which was logs and the rest was twigs. Twigs were mostly consumed in the villages, and logs were more popular in cities of India. The overall contribution of fuelwood, including sawdust and wood waste, was about 46 per cent of the total, the rest being agri waste and biomass dung cakes. Traditional fuel (fuelwood, crop residue and dung cake) dominates domestic energy use in rural India and accounts for about 90 per cent of the total. In urban areas, this traditional fuel constitutes about 24 per cent of the total.

Fuel wood, agri waste and biomass cake burning releases over 165 million tonnes of combustion products into India's indoor and outdoor air every year. To place this volume of emission in context, the Environmental Protection Agency (EPA) of the United States estimates that fire wood smoke contributes over 420,000 tonnes of fine particles throughout the United States – mostly during the winter months.

United States consumes about one-tenth of fuelwood consumed by India, and mostly for fireplace and home heating purposes. EPA estimates that residential wood combustion in the USA accounts for 44 per cent of total organic matter emissions and 62 per cent of the PAH, which are probable human carcinogens and are of great concern to EPA. The fuelwood sourced residential wood smoke makes up over 50 per cent of the wintertime particle pollution problem in California. In 2010, the state of California had about the same number of vehicles as all of India.

India burns tenfold more fuelwood every year than the United States, the fuelwood quality in India is different than the dry firewood of the United States, and the Indian stoves in use are less efficient thereby producing more smoke and air pollutants per kilogram equivalent. India has less land area and less emission air space than the United States. In summary, the impact on indoor and outdoor air pollution by fuelwood and biomass cake burning is far worse in India.

A United Nations study finds firewood and biomass stoves can be made more efficient in India. Animal dung, now used in inefficient stoves, could be used to produce biogas, a cleaner fuel with higher utilization efficiency. In addition, an excellent fertilizer can be produced from the slurry from biogas plants. Switching to gaseous fuels would bring the greatest gains in terms of both thermal efficiency and reduction in air pollution, but would require more investment. A combination of technologies may be the best way forward.

Between 2001 and 2010, India has made progress in adding electrical power generation capacity, bringing electricity to rural areas, and reforming market to improve availability and distribution of liquified cleaner burning fuels in urban and rural area. Over the same period, scientific data collection and analysis show improvement in India's air quality, with some regions witnessing 30 to 65 per cent reduction in NOx, SOx and suspended particulate matter. Even at these lower levels, the emissions are higher than those recommended by the World Health Organization. Continued progress is necessary.

Scientific studies conclude biomass combustion in India is the country's dominant source of carbonaceous aerosols, emitting 0.25 teragram per year of black carbon into air, 0.94 teragram per year of organic matter, and 2.04 teragram per year of small particulates with diameter less than 2.5 microns. Biomass burning, as domestic fuel in India, accounts for about 3 times as much black carbon air pollution as all other sources combined, including vehicles and industrial sources.

Other sources of pollution in Indian cities are vehicles and emissions from industry. Until 1992, India protected its automobile industry using license raj. Many two wheel, three wheel and four wheel vehicles lacked catalytic converters. Per vehicle emissions were amongst the highest in the world. The refining of oil and supply of fuel was owned, regulated and run by the government; the fuel quality was lax. In 2005, India adopted emission standard of Bharat Stage IV for vehicles, which is equivalent to Euro IV European standards for vehicle emissions. Nevertheless, the old pre-2005 vehicles, and even pre-1992 vehicles are still on Indian streets.

## FUEL ADULTERATION

Some Indian taxis and auto-rickshaws run on adulterated fuel blends. Adulteration of gasoline and diesel with lower-priced fuels is common in South Asia, including India. Some adulterants increase emissions of harmful pollutants from vehicles, worsening urban air pollution. Financial incentives arising from differential taxes are generally the primary cause of fuel adulteration. In India and other developing countries, gasoline carries a much higher tax than diesel, which in turn is taxed more than kerosene meant as a cooking fuel, while some solvents and lubricants carry little or no tax. As fuel prices rise, the public transport driver cuts costs by blending the cheaper hydrocarbon into highly taxed hydrocarbon. The blending may be as much as 20-30 per cent. For a low wage driver, the adulteration can yield short term savings that are significant over the month. The consequences to long term air pollution, quality of life and effect on health are simply ignored. Also ignored are the reduced life of vehicle engine and higher maintenance costs, particularly if the taxi, auto-rickshaw or truck is being rented for a daily fee.

Adulterated fuel increases tailpipe emissions of hydrocarbons (HC), carbon monoxide (CO), oxides of nitrogen (NOx) and particulate matter (PM). Air toxin emissions — which fall into the category of unregulated emissions— of primary concern are benzene and polyaromatic hydrocarbons (PAHs), both well known carcinogens. Kerosene is more difficult to burn than gasoline; its addition results in higher levels of HC, CO and PM emissions even from catalyst-equipped cars. The higher sulfur level of kerosene is another issue. The permissible level of fuel sulfur in India, in 2002, was 0.25 per cent by weight as against 0.10 per cent for gasoline. The higher levels of sulfur can deactivate the catalyst. Once the catalyst becomes deactivated, the amount of pollution from the vehicle dramatically increases. Fuel adulteration is essentially an unintended consequence of tax policies and the attempt to control fuel prices, in the name of fairness. Air pollution is the ultimate result. This problem is not unique to India, but prevalent in many developing countries including those outside of south Asia. This problem is largely absent in economies that do not regulate the ability of fuel producers to innovate or price based on market demand.

## TRAFFIC CONGESTION

Traffic congestion is severe in India's cities and towns. Traffic congestion is caused for several reasons, some of which are: increase in number of vehicles per kilometer of available road, a lack of intra-city divided-lane highways and intra-city expressways networks, lack of inter-city expressways, traffic accidents and chaos from poor enforcement of traffic laws.

Traffic congestion reduces average traffic speed. At low speeds, scientific studies reveal, vehicles burn fuel inefficiently and pollute more per trip. For example, a study in the United States found that for the same trip, cars consumed more fuel and polluted more if the traffic was congested, than when traffic flowed freely. At average trip speeds between 20 to 40 kilometres per hour, the cars pollutant emission was twice as much as when the average speed was 55 to 75 kilometres per hour. At average trip speeds between 5 to 20 kilometres per hour, the cars pollutant emissions were 4 to 8 times as much as when the average speed was 55 to 70 kilometres per hour. Fuel efficiencies similarly were much worse with traffic congestion.

Traffic gridlock in Delhi and other India cities is extreme. The average trip speed on many Indian city roads is less than 20 kilometres per hour; a 10 kilometer trip can take 30 minutes, or more. At such speeds, vehicles in India emit air pollutants 4 to 8 times more than they would with less traffic congestion; Indian vehicles also consume a lot more carbon footprint fuel per trip, than they would if the traffic congestion was less.

In cities like Bangalore, around 50 per cent of children suffer from asthma.

## RECENT TRENDS IN INDIA'S AIR QUALITY

With the last 15 years of economic development and regulatory reforms, India has made progress in improving its air quality. The table presents the average emissions sampled at many locations, over time, and data analyzed by scientific methods, by multiple agencies, including The World Bank. For context and comparison, the table also includes average values for Sweden in 2008, observed and analyzed by same methods. Over 1995-2008, average nationwide levels of major air pollutants have dropped by between 25-45 per cent in India.

India's Central Pollution Control Board now routinely monitors four air pollutants namely sulphur dioxide ($SO_2$), oxides of nitrogen (NOx), suspended particulate matter (SPM) and respirable particulate matter (PM10). These are target air pollutants for regular monitoring at 308 operating stations in 115 cities/towns in 25 states and 4 Union Territories of India.

The monitoring of meteorological parameters such as wind speed and direction, relative humidity and temperature has also been integrated with the monitoring of air quality. The monitoring of these pollutants is carried out for 24 hours (4-hourly sampling for gaseous pollutants and 8-hourly sampling for particulate matter) with a frequency of twice a week, to yield 104 observations in a year.

For 2010, the key findings of India's central pollution control board are:

- Most Indian cities continue to violate India's and world air quality PM10 targets. Respirable particulate matter pollution remains a key challenge for India. Despite the general non-attainment, some cities showed far more improvement than others. A decreasing trend has been observed in PM10 levels in cities like Solapur and Ahmedabad over the last few years. This improvement may be due to local measures taken to reduce sulphur in diesel and stringent enforcement by Gujarat government.
- A decreasing trend has been observed in sulphur dioxide levels in residential areas of many cities such as Delhi, Mumbai, Lucknow, Bhopal during last few years. The decreasing trend in sulphur dioxide levels may be due to recently introduced clean fuel standards, and the increasing use of LPG as domestic fuel instead of coal or fuelwood, and the use of LPG instead of diesel in certain vehicles.
- A decreasing trend has been observed in nitrogen dioxide levels in residential areas of some cities such as Bhopal and Solapur during last few years. The decreasing trend in sulphur dioxide levels may be due to recently introduced vehicle emission standards, and the increasing use of LPG as domestic fuel instead of coal or fuelwood.
- Most Indian cities greatly exceed acceptable levels of suspended particulate matter. This may be because of refuse and biomass burning, vehicles, power plant emissions, industrial sources.
- The Indian air quality monitoring stations reported lower levels of PM10 and suspended particulate matter during monsoon months possibly due to wet deposition and air scrubbing by rainfall. Higher levels of particulates were observed during winter months possibly due to lower mixing heights and more calm conditions. In other words, India's air quality worsens in winter months, and improves with the onset of monsoon season.

For its 2008 annual report, Central Pollution Control Board used 346 operating Air Quality Monitoring Stations, covering 130 cities/ towns in 26 States and 4 Union Territories. With the weekly data collected and then averaged over the year, Central Pollution Control Board reported the following annual trends from 1998 to 2008:

- The average annual SOx and NOx emissions level and periodic violations in industrial areas of India were significantly and surprisingly lower than the emission and violations in residential areas of India.
- The 24-hour average PM10 and suspended particulate matter emissions and violations in almost all areas of India violated India's and WHO targets. The PM10 and suspended particulate matter concentrations, in industrial areas of India were, however, lower than those in residential areas of India. Residential areas of India were the source of over 90 per cent of the most serious and repeated violations in particulate air pollution.
- Of the four major Indian cities, air pollution was consistently worst in Delhi, every year over 5 year period (2004–2008). Kolkata was a close second, followed by Mumbai. Chennai air pollution was least of the four.
- The states of Kerala and Meghalaya, relative to other Indian states, experienced on average some of lowest air pollution levels. The cities of Thiruvananthapuram, Kottayam and Shillong, relative to other Indian cities and towns, experienced some of lowest air pollution levels.
- Air Quality data collected from the monitoring station at Taj Mahal, Agra since year 1991 to 2008, suggests that both particulate and acid rain pollutants at Taj Mahal have been declining over the years. The 2008 average annual air pollutant concentrations were between 38 to 67 per cent lower than those in 1991.

## SOLID WASTE POLLUTION

Trash and garbage is a common sight in urban and rural areas of India. It is a major source of pollution. Indian cities alone generate more than 100 million tons of solid waste a year. Street corners are piled with trash. Public places and sidewalks are despoiled with filth and litter, rivers and canals act as garbage dumps.

In part, India's garbage crisis is from rising consumption. India's waste problem also points to a stunning failure of governance. In 2000, India's Supreme Court directed all Indian cities to implement a comprehensive waste-management programme that would include household collection of segregated waste, recycling and composting.

These directions have simply been ignored. No major city runs a comprehensive programme of the kind envisioned by the Supreme Court. Indeed, forget waste segregation and recycling directive of the India's Supreme Court, the Organization for Economic Cooperation and Development estimates that up to 40 per cent of municipal waste in India remains simply uncollected. Even medical waste, theoretically controlled by stringent rules that require hospitals to operate incinerators, is routinely dumped with regular municipal garbage. A recent study found that about half of India's medical waste is improperly disposed of.

Municipalities in Indian cities and towns have waste collection employees. However, these are unionized government workers and their work performance is neither measured nor monitored.

Some of the few solid waste landfills India has, near its major cities, are overflowing and poorly managed. They have become significant sources of greenhouse emissions and breeding sites for disease vectors such as flies, mosquitoes, cockroaches, rats, and other pests.

In 2011, several Indian cities embarked on waste-to-energy projects of the type in use in Germany, Switzerland and Japan. For example, New Delhi is implementing two incinerator projects aimed at turning the city's trash problem into electricity resource. These plants are being welcomed for addressing the city's chronic problems of excess untreated waste and a shortage of electric power. They are also being welcomed by those who seek to prevent water pollution, hygiene problems, and eliminate rotting trash that produces potent greenhouse gas methane. The projects are being opposed by waste collection workers and local unions who fear changing technology may deprive them of their livelihood and way of life.

Along with waste-to-energy projects, some cities and towns such as Pune, Maharashtra are introducing competition and the privatization of solid waste collection, street cleaning operations and bio-mining to dispose the waste. A scientific study suggests public private partnership is, in Indian context, more useful in solid waste management. According to this study, government and municipal corporations must encourage PPP-based local management through collection, transport and segregation and disposal of solid waste.

## NOISE POLLUTION

The Supreme Court of India gave a significant verdict on noise pollution in 2005. Unnecessary honking of vehicles makes for a highdecibel level of noise in cities. The use of loudspeakers for political purposes and by temples and mosques make for noise pollution inresidential areas because using more speakers in an programmes the noise is increase. In January 2010, Government of India published norms of permissible noise levels in urban and rural areas.

## LAND POLLUTION

In March 2009, the issue of Uranium poisoning in Punjab came into light, caused by fly ash ponds of thermal power stations, which reportedly lead to severe birth defects in children in the Faridkot and Bhatinda districts of Punjab.

## GREENHOUSE GAS EMISSIONS

India was the third largest emitter of carbon dioxide in 2009 at 1.65 Gt per year, after China (6.9 Gt per year) and the United States (5.2 Gt per year). With 17 per cent of world population, India contributed some 5 per cent of human-sourced carbon dioxide emission; compared to China's 24 per cent share. On per capita basis, India emitted about 1.4 tons of carbon dioxide per person, in comparison to the United States' 17 tons per person, and a world average of 5.3 tons per person.

About 65 per cent of India's carbon dioxide emissions in 2009 was from heating, domestic uses and power sector. About 9 per cent of India's emissions were from transportation (cars, trains, two wheelers, airplanes, others). India's coal-fired, oil-fired and natural gas-fired thermal power plants are inefficient and offer significant potential for $CO_2$ emission reduction through better technology.

Compared to the average emissions from coal-fired, oil-fired and natural gas-fired thermal power plants in European Union (EU-27) countries, India's thermal power plants emit 50 to 120 per cent more CO2 per kWh produced. This is in significant part to inefficient thermal power plants installed in India prior to its economic liberalization in the 1990s.

Between 1990 and 2009, India's carbon dioxide emissions per GDP purchasing power parity basis have decreased by over 10 per cent, a trend similar to China. Meanwhile, between 1990 and 2009, Russia's carbon dioxide emissions per GDP purchasing power parity basis have increased by 40 per cent. India has one of the better records in the world, of an economy that is growing efficiently on CO2 emissions basis. In other words, over the last 20 years, India has reduced CO2 emissions with each unit of GDP increase. Per Copenhagen Accord, India aims to further reduce emissions intensity of its growing GDP by 20 to 25 per cent before 2020, with technology transfer and international cooperation. Nevertheless, it is expected, that like China, India's absolute carbon dioxide emissions will rise in years ahead, even as International Energy Agency's Annex I countries expect their absolute CO2 emissions to drop.

A significant source of greenhouse gas emissions from India is from black carbon, NOx, methane and other air pollutants. These pollutants are emitted in large quantities in India every day from incomplete and inefficient combustion of biomass (fuel wood, crop waste and cattle dung). A majority of Indian population lacks access to clean burning fuels, and uses biomass combustion as cooking fuel. India's poorly managed solid wastes, inadequate sewage treatment plants, water pollution and agriculture are other sources of greenhouse gas emissions.

NASA's Lau has proposed that as the aerosol particles rise on the warm, convecting air, they produce more rain over northern India and the Himalayan foothill, which further warms the atmosphere and fuels a “heat pump” that draws yet more warm air to the region. This phenomenon, Lau believes, changes the timing and intensity of the monsoon, effectively transferring heat from the low-lying lands over the subcontinent to the atmosphere over the Tibetan Plateau, which in turn warms the high-altitude land surface and hastens glacial retreat. His modeling shows that aerosols—particularly black carbon and dust—likely cause as much of the glacial retreat in the region as greenhouse gases via this “heat pump” effect.

## IRRIGATION WATER MANAGEMENT

An adequate water supply is important for plant growth. When rainfall is not sufficient, the plants must receive additional water from irrigation. Various methods can be used to supply irrigation water to the plants. Each method has its advantages and disadvantages. These should be taken into account when choosing the method which is best suited to the local circumstances.

## IRRIGATION AND ARTIFICIAL APPLICATION OF WATER

Irrigation may be defined as the science of artificial application of water to the land or soil. It is used to assist in the growing of agricultural crops, maintenance of landscapes, and revegetation of disturbed soils in dry areas and during periods of inadequate rainfall. Additionally, irrigation also has a few other uses in crop production, which include protecting plants against frost, suppressing weed growing in grain fields and helping in preventing soil consolidation. In contrast, agriculture that relies only on direct rainfall is referred to as rain-fed or dryland farming. Irrigation systems are also used for dust suppression, disposal of sewage, and in mining. Irrigation is often studied together with drainage, which is the natural or artificial removal of surface and sub-surface water from a given area. Irrigation is also a term used in medical/dental fields to refer to flushing and washing out anything with water or another liquid.

## VARIOUS TYPES OF IRRIGATION

Various types of irrigation techniques differ in how the water obtained from the source is distributed within the field. In general, the goal is to supply the entire field uniformly with water, so that each plant has the amount of water it needs, neither too much nor too little.The modern methods are efficient enough to achieve this goal.

### SURFACE IRRIGATION SYSTEMS

In surface irrigation systems, water moves over and across the land by simple gravity flow in order to wet it and to infiltrate into the soil. Surface irrigation can be subdivided into furrow, *borderstrip or basin irrigation*. It is often called flood irrigation when the irrigation results in flooding or near flooding of the cultivated land. Historically, this has been the most common method of irrigating agricultural land. Where water levels from the irrigation source permit, the levels are controlled by dikes, usually plugged by soil. This is often seen in terraced rice fields (rice paddies), where the method is used to flood or control the level of water in each distinct field. In some cases, the water is pumped, or lifted by human or animal power to the level of the land.

**Localized**

Localized irrigation is a system where water is distributed under low pressure through a piped network, in a pre-determined pattern, and applied as a small discharge to each plant or adjacent to it. Drip irrigation, spray or micro-sprinkler irrigation and bubbler irrigation belong to this category of irrigation methods.

**Drip**

Drip irrigation, also known as trickle irrigation, functions as its name suggests.In this system water falls drop by drop just at the position of roots. Water is delivered at or near the root zone of plants, drop by drop. This method can be the most water-efficient method of irrigation, if managed properly, since evaporation and run-off are minimized. In modern agriculture, drip irrigation is often combined with plastic mulch, further reducing evaporation, and is also the means of delivery of fertilizer. The process is known as *fertigation*. Deep percolation, where water moves below the root zone, can occur if a drip system is operated for too long or if the delivery rate is too high. Drip irrigation methods range from very high-tech and computerized to low-tech and labour-intensive. Lower water pressures are usually needed than for most other types of systems, with the exception of low energy centre pivot systems and surface irrigation systems, and the system can be designed for uniformity throughout a field or for precise water delivery to individual plants in a landscape containing a mix of plant species.

Although it is difficult to regulate pressure on steep slopes, pressure compensating emitters are available, so the field does not have to be level. High-tech solutions involve precisely calibrated emitters located along lines of tubing that extend from a computerized set of valves.

**Sprinkler**

In sprinkler or overhead irrigation, water is piped to one or more central locations within the field and distributed by overhead high-pressure sprinklers or guns. A system utilizing sprinklers, sprays, or guns mounted overhead on permanently installed risers is often referred to as a *solid-set* irrigation system.

Higher pressure sprinklers that rotate are called *rotors* and are driven by a ball drive, gear drive, or impact mechanism. Rotors can be designed to rotate in a full or partial circle.

Guns are similar to rotors, except that they generally operate at very high pressures of 40 to 130 lbf/in$^2$ (275 to 900 kPa) and flows of 50 to 1200 US gal/min (3 to 76 L/s), usually with nozzle diameters in the range of 0.5 to 1.9 inches (10 to 50 mm). Guns are used not only for irrigation, but also for industrial applications such as dust suppression and logging.

Sprinklers can also be mounted on moving platforms connected to the water source by a hose. Automatically moving wheeled systems known as *traveling sprinklers* may irrigate areas such as small farms, sports fields, parks, pastures, and cemeteries unattended. Most of these utilize a length of polyethylene tubing wound on a steel drum.

As the tubing is wound on the drum powered by the irrigation water or a small gas engine, the sprinkler is pulled across the field. When the sprinkler arrives back at the reel the system shuts off. This type of system is known to most people as a "waterreel" traveling irrigation sprinkler and they are used extensively for dust suppression, irrigation, and land application of waste water. Other travellers use a flat rubber hose that is dragged along behind while the sprinkler platform is pulled by a cable. These cable-type travellers are definitely old technology and their use is limited in today's modern irrigation projects.

**Centre Pivot**

Centre pivot irrigation is a form of sprinkler irrigation consisting of several segments of pipe (usually galvanized steel or aluminum) joined together and supported by trusses, mounted on wheeled towers with sprinklers positioned along its length. The system moves in a circular pattern and is fed with water from the pivot point at the centre of the arc. These systems are found and used in all parts of the world and allow irrigation of all types of terrain. Newer systems have drop sprinkler heads as shown in the image that follows.

Most centre pivot systems now have drops hanging from a u-shaped pipe attached at the top of the pipe with sprinkler heads that are positioned a few feet (at most) above the crop, thus limiting evaporative losses. Drops can also be used with drag hoses or bubblers that deposit the water directly on the ground between crops. Crops are often planted in a circle to conform to the centre pivot. This type of system is known as LEPA (Low Energy Precision Application). Originally, most centre pivots were water powered. These were replaced by hydraulic systems (*T-L Irrigation*) and electric motor driven systems (Reinke, Valley, Zimmatic). Many modern pivots feature GPS devices.

**Lateral Move (Side Roll, Wheel Line)**

A series of pipes, each with a wheel of about 1.5 m diameter permanently affixed to its midpoint and sprinklers along its length, are coupled together at one edge of a field. Water is supplied at one end using a large hose. After sufficient water has been applied, the hose is removed and the remaining assembly rotated either by hand or with a purpose-built mechanism, so that the sprinklers move 10 m across the field. The hose is reconnected.

The process is repeated until the opposite edge of the field is reached. This system is less expensive to install than a centre pivot, but much more labour intensive to operate, and it is limited in the amount of water it can carry.

Most systems utilize 4 or 5-inch (130 mm) diameter aluminum pipe. One feature of a lateral move system is that it consists of sections that can be easily disconnected. They are most often used for small or oddly shaped fields, such as those found in hilly or mountainous regions, or in regions where labour is inexpensive.

Subirrigation also sometimes called *seepage irrigation* has been used for many years in field crops in areas with high water tables. It is a method of artificially raising the water table to allow the soil to be moistened from below the plants' root zone. Often those systems are located on permanent grasslands in lowlands or river valleys and combined with drainage infrastructure. A system of pumping stations, canals, weirs and gates allows it to increase or decrease the water level in a network of ditches and thereby control the water table.

Sub-irrigation is also used in commercial greenhouse production, usually for potted plants. Water is delivered from below, absorbed upwards, and the excess collected for recycling. Typically, a solution of water and nutrients floods a container or flows through a trough for a short period of time, 10–20 minutes, and is then pumped back into a holding tank for reuse. Sub-irrigation in greenhouses requires fairly sophisticated, expensive equipment and management. Advantages are water and nutrient conservation, and labour-saving through lowered system maintenance and automation. It is similar in principle and action to subsurface drip irrigation.

### Manual using Buckets or Watering Cans

These systems have low requirements for infrastructure and technical equipment but need high labour inputs. Irrigation using watering cans is to be found for example in peri-urban agriculture around large cities in some African countries.

### Automatic, Non-electric using Buckets and Ropes

Besides the common manual watering by bucket, an automated, natural version of this also exist. Using plain polyester ropes combined with a prepared ground mixture can be used to water plants from a vessel filled with water.

The ground mixture would need to be made depending on the plant itself, yet would mostly consist of black potting soil, vermiculite and perlite. This system would (with certain crops) allow to save expenses as it does not consume any electricity and only little water (unlike sprinklers, water timers...). However, it may only be used with certain crops (probably mostly larger crops that do not need a humid environment; perhaps *e.g.* paprikas).

### Using Water Condensed from Humid Air

In countries where at night, humid air sweeps the countryside, water can be obtained from the humid air by condensation onto cold surfaces. This is for example practiced in the vineyards at Lanzarote using stones to condense water or with various fog collectors based on canvas or foil sheets.

## IRRIGATION WATER QUALITY CRITERIA

Salt-affected soils develop from a wide range of factors including: soil type, field slope and drainage, irrigation system type and management, fertilizer and manuring practices, and other soil and water management practices. In Colorado, perhaps the most critical factor in predicting, managing, and reducing salt-affected soils is the quality of irrigation water being used. Besides affecting crop yield and soil physical conditions, irrigation water quality can affect fertility needs, irrigation system performance and longevity, and how the water can be applied. Therefore, knowledge of irrigation water quality is critical to understanding what management changes are necessary for long-term productivity.

### IRRIGATION WATER QUALITY CRITERIA

*Soil scientists use the following categories to describe irrigation water effects on crop production and soil quality*:

- Salinity hazard - total soluble salt content
- Sodium hazard - relative proportion of sodium to calcium and magnesium ions
- pH - acid or basic
- Alkalinity - carbonate and bicarbonate
- *Specific ions*: chloride, sulfate, boron, and nitrate.

Another potential irrigation water quality impairment that may affect suitability for cropping systems is microbial pathogens.

**Table. General Guidelines for Salinity Hazard of Irrigation Water based Upon Conductivity.**

| *Limitations for use* | *Electrical Conductivity* |
|---|---|
| | (dS/m)* |
| None | £0.75 |
| Some | 0.76 – 1.5 |
| Moderate | 1.51 – 3.00 |
| Severe | £3.00 |

*Note*: *dS/m at 25ºC = mmhos/cmLeaching required at higher range.Good drainage needed and sensitive plants may have difficulty at germination.

## SALINITY HAZARD

The most influential water quality guideline on crop productivity is the water salinity hazard as measured by electrical conductivity ($EC_w$). The primary effect of high $EC_w$ water on crop productivity is the inability of the plant to compete with ions in the soil solution for water (physiological drought). The higher the EC, the less water is available to plants, even though the soil may appear wet. Because plants can only transpire "pure" water, usable plant water in the soil solution decreases dramatically as EC increases. Actual yield reductions from irrigating with high EC water varies substantially. Factors influencing yield reductions include soil type, drainage, salt type, irrigation system and management.

**Table. Potential Yield Reduction from Saline Water for Selected Irrigated Crops.**

| | *% Yield Reduction* | | | |
|---|---|---|---|---|
| *Crop* | *0%* | *10%* | *25%* | *50%* |
| | | $EC_w$ | | |
| Barley | 5.3 | 6.7 | 8.7 | 12 |
| Wheat | 4.0 | 4.9 | 6.4 | 8.7 |
| Sugarbeet | 4.7 | 5.8 | 7.5 | 10 |
| Alfalfa | 1.3 | 2.2 | 3.6 | 5.9 |
| Potato | 1.1 | 1.7 | 2.5 | 3.9 |
| Corn (grain) | 1.1 | 1.7 | 2.5 | 3.9 |
| Corn (silage) | 1.2 | 2.1 | 3.5 | 5.7 |
| Onion | 0.8 | 1.2 | 1.8 | 2.9 |
| Dry Beans | 0.7 | 1.0 | 1.5 | 2.4 |

*Note*:

$EC_w$ = electrical conductivity of the irrigation water in dS/m at 25oC. Sensitive during germination. $EC_w$ should not exceed 3 dS/m for garden beets and sugarbeets.

The amount of water transpired through a crop is directly related to yield; therefore, irrigation water with high $EC_w$ reduces yield potential (Table). Beyond effects on the immediate crop is the long term impact of salt loading through the irrigation water. Water with an $EC_w$ of only 1.15 dS/m contains approximately 2,000 pounds of salt for every acre foot of water. You can use conversion factors in Table to make this calculation for other water EC levels.

Other terms that laboratories and literature sources use to report salinity hazard are: salts, salinity, electrical conductivity ($EC_w$), or total dissolved solids (TDS). These terms are all comparable and all quantify the amount of dissolved "salts" (or ions, charged particles) in a water sample. However, TDS is a direct measurement of dissolved ions and EC is an indirect measurement of ions by an electrode.

Although people frequently confuse the term "salinity" with common table salt or sodium chloride (NaCl), EC measures salinity from all the ions dissolved in a sample. This includes negatively charged ions (*e.g.*, Cl, $NO_3$) and positively charged ions (*e.g.*, Ca, Na). Another common source of confusion is the variety of unit systems used with $EC_w$. The preferred unit is deciSiemens per meter (dS/m), however millimhos per centimeter (mmho/cm) and micromhos per centimeter ($\mu$mho/cm) are still frequently used.

**Table. Conversion Factors for Irrigation Water Quality Laboratory Reports.**

| Component | To Convert | Multiply By | To Obtain |
|---|---|---|---|
| Water nutrient or TDS | mg/L | 1.0 | ppm |
| Water salinity hazard | 1 dS/m | 1.0 | 1 mmho/cm |
| Water salinity hazard | 1 mmho/cm | 1,000 | 1 μmho/cm |
| Water salinity hazard | $EC_w$ (dS/m) for EC <5 dS/m | 640 | TDS (mg/L) |
| Water salinity hazard | $EC_w$ (dS/m) for EC >5 dS/m | 800 | TDS (mg/L) |
| Water $NO_3N$, $SO_4$-S,B applied | ppm | 0.23 | lb per acre inch of water |
| Irrigation water | acre inch | 27,150 | gallons of water |

## DEFINITIONS

| *Abbrev.* | *Meaning* |
|---|---|
| mg/L | Milligrams per litre |
| meq/L | Milliequivalents per litre |
| ppm | Parts per million |
| dS/m | DeciSiemens per meter |
| μS/cm | MicroSiemens per centimeter |
| mmho/cm | Millimhos per centimeter |
| TDS | Total dissolved solids |

## SODIUM HAZARD

### Infiltration/Permeability Problems

Although plant growth is primarily limited by the salinity ($EC_w$) level of the irrigation water, the application of water with a sodium imbalance can further reduce yield under certain soil texture conditions. Reductions in water infiltration can occur when irrigation water contains high sodium relative to the calcium and magnesium contents. This condition, termed "sodicity," results from excessive soil accumulation of sodium. Sodic water is not the same as saline water. Sodicity causes swelling and dispersion of soil clays, surface crusting and pore plugging. This degraded soil structure condition in turn obstructs infiltration and may increase run-off. Sodicity causes a decrease in the downward movement of water into and through the soil, and actively growing plants roots may not get adequate water, despite pooling of water on the soil surface after irrigation.

The most common measure to assess sodicity in water and soil is called the Sodium Adsorption Ratio (SAR). The SAR defines sodicity in terms of the relative concentration of sodium (Na) compared to the sum of calcium (Ca) and magnesium (Mg) ions in a sample. The SAR assesses the potential for infiltration problems due to a sodium imbalance in irrigation water. The SAR is mathematically written below, where Na, Ca and Mg are the concentrations of these ions in milliequivalents per litre (meq/L). Concentrations of these ions in water samples are typically provided in milligrams per litre (mg/L). To convert Na, Ca, and Mg from mg/L to meq/L, you should divide the concentration by 22.9, 20, and 12.15 respectively. For most irrigation waters encountered in Colorado the standard SAR formula provided above is suitable to express the potential sodium hazard. However, for irrigation water with high bicarbonate ($HCO_3$) content, an "adjusted" SAR ($SAR_{ADJ}$) can be calculated. In this case, the amount of calcium is adjusted for the water's alkalinity, is recommended in place of the standard SAR. Your laboratory may calculate an adjusted SAR in situations where the $HCO_3$ is greater than 200 mg/L or pH is greater than 8.5.

$$SAR = \frac{Na^{+} meq / L}{\sqrt{\frac{\left(Ca^{++}{}_{meq/L}\right) + \left(Mg^{++}{}_{meq/L}\right)}{2}}}$$

meq/L = mg/L divided by atomic weight of ion divided by ionic charge (Na = 23.0 mg/meq, Ca = 20.0 mg/meq, Mg = 12.15 mg/meq) The potential soil infiltration and permeability problems created from applications of irrigation water with high "sodicity" cannot be adequately assessed on the basis of the SAR alone. This is because the swelling potential of low salinity ($EC_w$) water is greater than high $EC_w$ waters at the same sodium content. Therefore, a more accurate evaluation of the infiltration/permeability hazard requires using the electrical conductivity ($EC_w$) together with the SAR.

**Table. Guidelines for Assessment of Sodium Hazard of Irrigation Water Based on SAR and $EC_w$.**

| | *Potential for Water Infiltration Problem* | |
|---|---|---|
| *Irrigation water SAR* | *Unlikely* | *Likely* |
| | ------ $EC_w$ (dS/m) ------ | |
| 0-3 | >0.7 | <0.2 |
| 3-6 | >1.2 | <0.4 |
| 6-12 | >1.9 | <0.5. |
| 12-20 | >2.9 | <1.0 |
| 20-40 | >5.0 | <3.0 |

Many factors including soil texture, organic matter, cropping system, irrigation system and management affect how sodium in irrigation water affects soils. Soils most likely to show reduced infiltration and crusting from water with elevated SAR (greater than 6) are those containing more than 30 per cent expansive (smectite) clay. Soils containing more than 30 per cent clay include most soils in the clay loam, silty clay loam textural classes and finer and some sandy clay loams. In Colorado, smectite clays are common in areas with agricultural production.

**Table. Susceptibility Ranges for Crops to Foliar Injury from Saline Sprinkler Water.**

| | Na or Cl Concentration (mg/L) Causing Foliar Injury | | | |
|---|---|---|---|---|
| Na concentration | <46 | 46-230 | 231-460 | >460 |
| Cl concentration | <175 | 175-350 | 351-700 | >700 |
| | Apricot | Pepper | Alfalfa | Sugarbeet |
| | Plum | Potato | Barley | Sunflower |
| | Tomato | Corn | Sorghum | |

## pH and Alkalinity

The acidity or basicity of irrigation water is expressed as pH (< 7.0 acidic; > 7.0 basic). The normal pH range for irrigation water is from 6.5 to 8.4. Abnormally low pH's are not common in Colorado, but may cause accelerated irrigation system corrosion where they occur. High pH's above 8.5 are often caused by high bicarbonate ($HCO_3$) and carbonate ($CO_3$) concentrations, known as alkalinity. High carbonates cause calcium and magnesium ions to form insoluble minerals leaving sodium as the dominant ion in solution. As described in the sodium hazard section, this alkaline water could intensify the impact of high SAR water on sodic soil conditions. Excessive bicarbonate concentrates can also be problematic for drip or micro-spray irrigation systems when calcite or scale build up causes reduced flow rates through orifices or emitters. In these situations, correction by injecting sulfuric or other acidic materials into the system may be required.

## Chloride

Chloride is a common ion in Colorado irrigation waters. Although chloride is essential to plants in very low amounts, it can cause toxicity to sensitive crops at high concentrations. Like sodium, high chloride concentrations cause more problems when applied with sprinkler irrigation. Leaf burn under sprinkler from both sodium and chloride can be reduced by night time irrigation or application on cool, cloudy days. Drop nozzles and drag hoses are also recommended when applying any saline irrigation water through a sprinkler system to avoid direct contact with leaf surfaces.

**Table. Chloride Classification of Irrigation Water.**

| *Chloride (ppm)* | *Effect on Crops* |
|---|---|
| Below 70 | Generally safe for all plants. |
| 70-140 | Sensitive plants show injury. |
| 141-350 | Moderately tolerant plants show injury. |
| Above 350 | Can cause severe problems. |

**Boron**

Boron is another element that is essential in low amounts, but toxic at higher concentrations. In fact, toxicity can occur on sensitive crops at concentrations less than 1.0 ppm. Colorado soils and irrigation waters contain enough B that additional B fertilizer is not required in most situations. Because B toxicity can occur at such low concentrations, an irrigation water analysis is advised for ground water before applying additional B to crops.

**Table. Boron Sensitivity of Selected Colorado Plants (B Concentration, mg/ L*)**

| Sensitive | Moderately Sensitive | Moderately Tolerant | Tolerant |
|---|---|---|---|
| 0.76-1.0 | 1.1-2.0 | 2.1-4.0 | 4.1-6.0 |
| Wheat | Carrot | Lettuce | Alfalfa |
| Barley | Potato | Cabbage | Sugar beet |
| Sunflower | Cucumber | Corn | Tomato |
| Dry Bean | | Oats | |

**Sulfate**

The sulfate ion is a major contributor to salinity in many of Colorado irrigation waters. As with boron, sulfate in irrigation water has fertility benefits, and irrigation water in Colorado often has enough sulfate for maximum production for most crops. Exceptions are sandy fields with <1 per cent organic matter and <10 ppm SO-S in irrigation water.

**Nitrogen**

Nitrogen in irrigation water (N) is largely a fertility issue, and nitrate-nitrogen ($NO_3$-N) can be a significant N source in the South Platte, San Luis Valley, and parts of the Arkansas River basins. The nitrate ion often occurs at higher concentrations than ammonium in irrigation water. Waters high in N can cause quality problems in crops such as barley and sugar beets and excessive vegetative growth in some vegetables. However, these problems can usually be overcome by good fertilizer and irrigation management. Regardless of the crop, nitrate should be credited towards the fertilizer rate especially when the concentration exceeds 10 ppm $NO_3$-N (45 ppm $NO_3^-$).

## IMPORTANCE OF IRRIGATION SYSTEM

John Skutsch and Darren Evans have made a valuable contribution to the debate on irrigation maintenance. They suggest that widespread evidence points to a considerable shortfall between recommended maintenance expenditure for public-sector irrigation schemes and the amounts actually spent. In a variety of projects researched during the 1990s in India, Indonesia, Pakistan and Sri Lanka, the ratio per hectare of actual spend to recommended spend was only

24-47 per cent. In the first years of a new scheme, maintenance neglect may have no dramatic effect.

Irrigation staff can freeride on the freeboard produced by the vertical distance by which the channel flow's usual height exceeds the required operating level. Cropping systems, too, may show some resilience. But sooner or later maintenance underfunding begins to bite. Schemes in arid and semi-arid areas, where crops are entirely dependent on irrigation under managed rotations, are more sensitive to neglect than projects in the humid tropics that have paddy rice under continuous irrigation supplementary to monsoon rains.

The biggest losses come through crop output decline as the volume of irrigation water delivered eventually begins to decline. This may show itself as a fall in tonnage efficiency or in a diminution of the irrigated area. Moreover, as Carruthers and Morrison indicate, farmers shift to lower value crops so as to be able to reduce risk by limiting the use of inputs. Impeded drainage can lead to waterlogging and salinity and to an increase in the incidence of water-related diseases associated with blocked channels and stagnant water. Skutsch and Evans, commenting on the serious social and financial implications for farmers, write:

Those in the more favoured parts of the system may continue to receive an adequate water supply, while those at the tail-end can face ruin. Poor maintenance thus directly aggravates existing inequities within the farming community. It initiates a vicious circle of decline: reduced water supply; lost output; farmers' anger, despair and reduced investment; reduced water fee collections; vandalism and conflict. Smallholder farmers operate on a narrow margin between relative success and failure. Land is often mortgaged against the following harvest to pay for the cost of agricultural inputs. A single disastrous cropping season can mean the loss of a farmer's land and enforced migration to the cities to seek work.

This process brings with it the premature obsolescence of irrigation projects. The search for new capital funding begins in order to renovate the assets far earlier than would have been necessary under a satisfactory maintenance regime.

A cycle of build-neglect-rebuild is in place. Partly as a result of this, most development bank finance for irrigation is now for the rehabilitation of existing schemes. Skutsch estimates that about two-thirds of recent international lending has been for systems that have suffered early technical failure. Already in 1995 Jones's overview of World Bank lending for irrigation projects could conclude: 'O&M problems can be seen in the Bank's financing of so many rehabilitation projects. Almost all of them, when scrutinized, turn out to be deferred maintenance projects.

Why, then, is maintenance neglected? As so often is the case in the social sciences, the question is simple but the answer is fiendishly complex. In part, farmers themselves are responsible. As is commonly the case with public-sector irrigation schemes, they see maintenance as the proper task of government.

If farmers are required to allocate greater resources to maintenance, either in terms of increased payments or of increased responsibility...they must see a clear incentive in better cash returns, reduced costs or improved control over water. Particularly on rice-growing schemes in monsoon climates, farmers may see little urgency in paying today to avoid increasing problems tomorrow, particularly if the government has traditionally stepped in to rehabilitate defective systems. In part, government and irrigation department staff are responsible for the neglect of maintenance.

Too often, national, regional and local élites regard the public sector as a milch cow whose teats are there to be squeezed for easy profits. No simpler way exists of doing this than by pushing superfluous staff into administrative posts as a favour to family, professional or political allies. As a consequence the 'establishment' of the department becomes excessively large, draining resources from OM&M. Moreover, operations activities in any case enjoy precedence over maintenance.

Technical and engineering staff, trained in design and construction, are likely to be unmotivated by the bread-and-butter tasks of OM&M, which offer poor rewards and prospects. For senior staff, administrators and engineers, new projects are likely to be the most highly sought after given their prestige and the rent-skimming opportunities they offer. A rule of thumb of some international consultants in parts of the developing world is that capital rather than current spending is favoured by country élites in government because 40 per cent of capital costs go in pay-offs.

At a level more mundane but of real importance, irrigation departments commonly remit the charges they collect from farmers to the finance ministry. That ministry may give other sectors more support than agriculture when it provides revenue financing. As the sum allocated to OM&M is decoupled from the fees collected from farmers, irrigation agencies have no incentive to maximize their revenue collection. There is a parallel here with the breaking of the line in prudential responsibility, itself a 'decoupling' of loan provider and loan user.

In part, the development banks themselves are responsible for maintenance neglect. If they lend money on schemes that 15 years later are choked with weeds, silted up, waterlogged, saline, exhibiting high distribution losses and with defective equipment, then their original cost-benefit analyses must have been *seriously* erroneous. The McNamara effect and a weak record in post-project evaluation are partially to blame here. The development banks need not fear that their loans will never be repaid, because the dollar flow of principal and interest comes from national government treasuries, not from the deficient project.

*Several authors have proposed policy changes aimed at ending the buildneglect-rebuild syndrome. Such proposals include the following*:

- An information system should be put in place that is capable of defining in detail the volume of water used.
- A dependable delivery system should be achieved.
- An agency willing and able to collect fees should be built, with a transparent management system.
- Farmers should be required to contribute an agreed number of maintenance days each year.
- Fees paid by farmers should increase but only after
  - The managing agency details all costs, levels of service and benefits,
  - Users are surveyed on their willingness and ability to pay,
  - The charging mechanism is agreed,
  - Users participate in setting the levels of service
  - Fees are linked to the level of service.
- Recognized sanctions for non-payment should be enforced.
- Feeding back the charges paid by farmers directly to OM&M.
- Maintenance should be carried out by private or autonomous self-accounting agencies.
- Engineers and planners should design irrigation projects for low maintenance, less costly management and greater effective lives.
- Improvement and modernization projects would be considered only for those schemes where a regular review of their individual maintenance standard shows them to be satisfactory or better.
- Lending agencies should provide financial support for a transitional period following construction.
- Tradeable abstraction rights should be introduced.
- Irrigation management should be transferred to farmers themselves.

## MANAGING COMMON PROPERTY RESOURCES

The above list of thirteen proposals for ending the build-neglect-rebuild cycle ends with irrigation management transfer from the public sector to farmers themselves. In the 1990s of such a possibility was able to draw upon and contribute to, a vigorous literature exploring the subject of common property resource management. The published papers and books demonstrate a welcome (if rare) willingness of economists and sociologists to work together in common cause. Below, synthesis of the relevant literature, particularly as it applies to the management of irrigation systems.

In every society common property resources exist, each of which is the common location of work by human agents or actors-the two terms are deployed here interchangeably. Agents may be individual persons, families or institutions.

Such common property resources may be constituted by specific natural environments, such as a commons used for grazing livestock, or by some complex resource integrating features of the natural environment with the means of production created by human society, such as an irrigation scheme. The 'common property' characteristic of the resource is that a number of separate actors enjoy rights of access and use of the resource-rights that are recognized in law or are customary practice.

In the day-to-day work located on the common resource, each agent may be motivated only by the *private* interest of that person, family or institution. This is of special importance in three ways. *First,* each actor may seek to appropriate as much of the resource for itself, without regard for other agents' interests. *Second,* each actor may choose to maintain the productivity of the resource only to the degree that the private costs borne by the actor in such efforts are outweighed by the private advantages, without regard for other agents' interests. *Third,* insofar as the use of the common resource creates negative side-effects, the actor may seek to reduce these side-effects only to the extent that their cost outweighs their production advantage, without regard for other agents' interests. So, for any single actor, private interest may prevail over the common interest in respect of resource appropriation, resource maintenance and resource degradation.

Production of outputs from a common resource-the milk yield of ruminants or the tonnage of grain-may grow vigorously over time. Work motivated by the private interests of the agent can stimulate long hours in the field, a sharp tactical appreciation of gains and losses from adaptive action and an eager search for innovative practices. In respect of the productivity of the common resource, these are powerful advantages.

However, with the passage of time the common property resource itself and the patterns of productive activity on it may show signs of enfeeblement and impending long-term collapse. Now, the disadvantages imposed by each single agent's actions on other persons, families and institutions have become disproportionately large. The struggle by actors over the appropriation of the resource can weaken friendship, create mistrust and result in disabling and destructive legal and physical conflicts as well as the exhaustion of the resource itself. Maintenance and protection of the common resource may decline below what is required for its long-term sustainability. Each agent's indifference to the negative effects of its actions on others may bring with it both bitter social disputes as well as the poisoning of the common resource.

Such a trajectory may lead to demands for a reconstitution of the social relationships between actors in the management of the common property resource. Where government cannot or will not lead this collective activity, a non-governmental local entity for collective action (LECA) may be created. The core objectives of the LECA are likely to be fourfold. *First,* the shared use of the resource should be recognized by the participant agents as being equitable. *Second,* the maintenance of the resource should be adequate to ensure its long-term viability. *Third,* the production of negative externalities should be sufficiently well regulated that their collective cost to the new institution's actors is acceptable and does not threaten systemic sustainability. *Fourth,* the transaction costs of meeting the first three objectives are acceptable to actors in the light of the benefits they bring.

There is a problem. A successful LECA brings all-round advantages to its actor-members. But every agent knows that if *he and he alone* continues to pursue his private interest, he reaps the benefits of unrestrained action as well as the benefits deriving from the collective agreement. This is known as the freerider problem. As freerider numbers grow, the advantages of collective action diminish and the LECA collapses.

So each LECA needs to engage in forms of moral persuasion in order that its actors honour the collective agreement made. It will also monitor members' activities to ensure that they do not breach the rules and that, when this occurs, sanctions are exercised against such infractions. The collective also faces costs of bargaining, contract formulation and information search in pursuit of a collective economic strategy. Together, all these costs of motivation, control and co-ordination are termed transaction costs. Even in a study as exhaustive as Anna Blomqvist's in South India it was never possible to assign these costs a monetary value.

The greater the LECA's legitimacy, *i.e.* the greater the internalization of its values and ideas within and outside the institution, the less costly is monitoring and enforcement. Blomqvist concluded from her experience that LECAs are most likely to succeed when the members are culturally homogeneous, small in number, highly dependent on the common resource and benefit equally from collective action.

## ABSTRACTION CHARGES AND SUSTAINABLE CATCHMENT MANAGEMENT

The subject of abstraction charging is related to the debate on sustainability, beginning with the approach of Richard Dubourg whose contributions to the study of hydroeconomics are within the neo-classical framework. Dubourg suggests that aggregate sustainability is a situation where natural capital as well as non-natural capital are non-declining and specifically where 'critical capital' such as water, within the natural capital category, is non-declining.

From these definitions he deduces impeccably that aggregate sustainability is consistent with catchment management policies in which abstraction of surface water and groundwater is equal to effective rainfall. It has to be said that the adoption of such a definition of sustainable abstraction would be extraordinarily dangerous.

Abstraction at a rate equal to effective rainfall may be capable of leaving surface and groundwater stocks unchanged over the course of a year so that the critical capital stock is maintained. But a rate of abstraction at this maximum rate would have the effect of capturing the entire river flow at some point or points in the catchment. In effect, Dubourg's definition ignores completely what in the USA is termed the environmental need for water in river systems-all on the basis of three mathematical constraints. In *Introduction to the Economics of Water Resources* I have suggested an alternative, multi-dimensional approach. After defining the concept of a 'sustainable society', it is suggested that water resource planning in such a society has six principal fields of action:

- The protection of water's hydrocyclical capacity to renew its ground and surface-water flows and stocks;
- The conservation of society's species and natural habitats in all their fresh and saltwater environments;
- The husbandry of water in its supply and use;
- The supply of freshwater sufficient to meet the biological, cultural and economic needs of society's human populations;
- The purification of water from domestic, agricultural and industrial effluents;
- The drainage of water and the protection of rural and urban communities against flood.

Abstraction charges can contribute positively to the first three of these fields and, below, it is shown how this can be done, using what I shall call *full cost incentive charging*.

The preparation of a tariff scheme should take as its starting point the regulatory controls put in place by the catchment authority in carrying out its responsibilities. In the development of these controls, there is an important role for social cost-effectiveness analysis and social cost-benefit analysis. These techniques have a place, alongside environmental impact assessment, in comparing alternative regulatory options. The design of environmental regulation should be an economic process as well as an ecological one.

The tariff regime introduced to any specific catchment (or region) will be contingent on its physical geography, its habitats and species, its human settlement patterns, its economy and the power relationships which hold between various social and economic groups. So what we require are criteria for tariff design which can be applied in the appropriate way to any single area with all its unique characteristics. I propose three such criteria for full cost incentive charging.

- *First,* the annual income from abstraction charges should be hypothecated to the environmental regulator such that, when added to the income received from discharge fees, fishing licences, navigation permits, etc., abstraction fee income is sufficient to cover all the state's capital and current account expenditures on environmental regulation, research and database development, compensation payments, etc. In providing the regulator with hypothecated income, it is likely to increase their relative power as an agency of government. As Kraemer writes of Germany since 1988: 'On the whole, the water resource taxes contributed to capacity building within the water management administration in the German Länder and thus partly overcame the implementation deficit in water resource management'. It will also provide a budget to finance the legal costs of modifying or terminating abstraction licences that threaten the hydrocycle or undermine nature conservation. Higher abstraction charges will also give a price signal to water companies to reduce their storage and distribution leakage between the points of abstraction and the user's gate.

- *Second,* where the full cost abstraction tariff still leaves an excess of demand for abstracted water over its licensed supply, the charge should be raised so that market clearing takes place.
- *Third,* specification of the components of the abstraction charge should provide incentives for abstraction behaviour that is economically efficient and that avoids environmental degradation. Price per unit volume should be invariant with total volume abstracted, unless there are countervailing economic or environmental arguments. Price should be discounted where abstractors recycle their off-take to surface or groundwater sources. (Discharge fees should be used to handle the water quality aspects of recycled water.) Charges should be higher for abstracted water of higher quality. Seasonal variations in effective rainfall and economic demand should be dealt with through the licensed volume provisions laid down by the regulator and by the market-clearing criterion. Charges should be higher for upstream sources and for abstraction in locations where species and habitats are more threatened by abstraction.

The institutional framework that would be most appropriate for full cost incentive charging deserves discussion and my ideas here have been strongly influenced by Karin Kemper's *The Cost of Free Water*. The basic approach is a negotiation model in which there exists a public-sector catchment agency that, through a negotiating forum, develops its policies with the advice of water companies, direct abstractors, environmental organizations and water user associations representing the domestic sector, irrigated agriculture, mining and manufacturing, etc.

The original prototype for such negotiation models is the French water parliaments. Central government retains the statutory right to determine which public and private institutions may enjoy the right to abstract water, on what scale, in what locations, at what time of year, at what price-but it delegates such rights to the catchment agency.

The agency, with the assistance of its partners in the negotiating forum and with a full understanding of existing customary rights, then assigns formal abstraction rights on a time-limited basis to specific abstractors or groups of abstractors. The time limit would be 10 years, rolled over each year except where the agency allows the licence to expire. Where these 10-year rights need to be modified or rescinded for hydrological, environmental or economic reasons, compensation would be payable to the abstractor so affected. Rules would also exist to address third party impacts. Abstraction rights assigned to abstractors could be freely traded provided the agency had approved such transfers in the light of their social, economic and environmental impacts. Full cost incentive charging would be the basis of the price paid by abstractors for their water.

*It has the objectives of*:

- Underpinning environmental regulation with a hypothecated income source;
- Requiring abstractors (and therefore final users) to cover the full costs of regulation;
- Bringing the quantity of water demanded by abstractors into line with regulatory limits; and
- Giving price signals that promote both allocative efficiency as well as abstraction practices that avoid damage to riverine eco-systems.

No charge would be made for abstractions below a minimum scale-the transaction costs of such charges would be high compared with the volumes abstracted.

Arrangements would be made to monitor abstraction with respect to its location, time and quantity, as well as to invoice abstractors, to collect the charges owed and to enforce all agreements. Such transaction costs would be included in the full costing tariff of abstraction charges. The creation of this institutional framework imposes social, economic and political costs on the parties concerned, structural costs of change both real and financial in their nature. Therefore the negotiating forum may agree that it is sensible for full cost incentive charging to be phased in gradually.

## Irrigation Management Transfer on Mexico

Local associations of water users serve the same functions as irrigation agencies, but on a very localized level. In some countries such as Nepal and the Philippines, the major portion of the irrigated area is managed locally, through village-based water user associations. Typically such associations manage very small irrigation canals that were constructed by the users, perhaps centuries ago and the associations have grown up around the need for operating and maintaining these canals. In many developing countries such traditional canal systems have been the target of modernisation efforts by government agencies, often funded through international development assistance. The relatively crude physical infrastructure of many of these canal systems was rebuilt and absorbed into the domain of the government agency, which replaced the management functions of the indigenous water user associations.

Today, local water user associations are recognised as resources of 'social capital' which will have an increasingly important role to play in the coming decades. Instead of absorbing such associations into the state organisation, the capacity of associations is developed so that they can improve the performance of their own irrigation and drainage systems. This participatory approach was re-invented in the Philippines in the 1970s and has since evolved into a global trend that combines participation with various degrees of privatisation. In many developing countries a major focus of this process is the institutional challenge of establishing new water user associations which can serve the management functions previously handled by government.

A particularly fruitful source for a case study of this trend is Mexico. With 5 million irrigated hectares it was the seventh largest irrigator in the world in 1995. It began the irrigation management transfer process as early as 1989 and the experience has been extensively researched. Finally, Mexico has for a number of years been considered as a paradigm for this form of institutional transformation. The case study draws heavily on the work of Kloezen, Garcés-Restrepo and Johnson.

The broad political and economic context within which Mexico introduced irrigation management transfer was central government's neo-liberal response to the economic crisis of the 1980s and the unsatisfactory performance of the irrigation districts under the management of the National Water Commission. Transfer has been a top-down process motivated by the international development banks, the main objective being to reduce public expenditure on irrigation OM&M whilst promoting greater user participation in irrigation management. It has been accompanied by a new National Water Act, revision of the Constitution to give legal foundation for the privatization of the *ejidos* (land reform communities) in all irrigation districts and termination of guaranteed crop prices and subsidized credit. 'Dismantling the public sector, including the public irrigation sector, would not have been possible without the commitment at the highest political levels to reduce staff working in the public sector'.

The institutional transformation has been rapid and radical. In its first phase the National Water Commission retained management of the headworks and the main canals. Water user associations took over financial and managerial responsibility for OM&M below the main canal. In the second phase the Commission is left with management of the reservoirs and surface-water-pumping stations; a District Federation of the user associations takes on the OM&M of the main canals. These associations are non-governmental local entities for collective action.

The high speed of the process and the low resistance of farmers have surprised many observers. It seems that transfer was preceded by other neo-liberal reforms in agriculture. Farmers knew the Commission's traditional OM&M was to be terminated and OM&M had in any case declined in quality. Reform came first in the larger districts in the north, where many large private producers were known to support change. Lastly, the programme built on a strong organizational base-the *ejidos* themselves and the private farmers' co-operatives and unions.

Government informed farmers of the impending change-there seems to have been no serious consultation-and instructed them to select their delegates to the new user associations. Thereafter, delegates to each association elected from among themselves a president, a treasurer and a secretary. The Commission worked with the new user groups for 6 months or more from the time of transfer and provided their leaders and technical staff with extensive training in OM&M and financial management. The Commission also granted concessions to the associations to use its machinery and equipment so that there would be no capital expenditure required in advance.

Prior to transfer the National Water Commission was wholly responsible at the district level for the planning of annual and seasonal water allocations. Similarly the Commission employed the heads of the irrigation units into which each district was divided. These units were more or less independent hydraulic blocks with sizes ranging from 3,000 to 20,000 ha. Unit heads and their channel-keepers were responsible for daily OM&M at all system levels down to the farm inlets. Farmers paid their fees at the unit office and were given water by the channel-keeper.

After transfer, hydraulic committees were introduced at district level for allocation planning; these committees were composed of representatives from the Commission, the state government and each user association in that district. At the same time the district 'units' were replaced by 'modules'-two or more per unit-and it is at this level that the new water user associations manage.

They collect irrigation water fees directly from farmers and employ their own module managers, channel-keepers, maintenance personnel and administrative staff. Based on the volume of water a module takes from the Commission and on the proportional amount of main infrastructure serving a module, each association must pay a percentage of the total fees it collects to the Commission.

The specific subject of this case study is the Alto Río Lerma Irrigation District. Located in the State of Guanajuato, it has an area of 113,000 ha. There are some 24,000 farmers of whom 55 per cent are from the *ejidos* and 45 per cent are private growers. The average landholding is 5 ha. Private holdings average twice the size of *ejido* holdings. The district's climate is sub-humid with annual precipitation of 750mm and potential evapotranspiration of 1900mm. Eighty millimetres of rain falls in the winter season from November to April and 670 mm falls from May to November. Average temperature is 19 °C and relative humidity is 60 per cent.

The Alto Río Lerma Irrigation District enjoys access to both surface water and groundwater. The river has four storage dams with a combined capacity of 2,140 mcm, as well as five diversion dams. The distribution network has 475 km of main canals and 1,660 km of secondary and tertiary canals, as well as 1,030 km of drainage canals. There are 1,710 deep wells exploiting three different aquifers with a total annual recharge of 500 mcm. Seventy per cent of the irrigated hectarage uses surface water and 30 per cent uses groundwater. Wheat and barley are the main crops in the dry winter months. Sorghum, maize and beans predominate in the wet summer season. All farmers grow vegetables and the private growers do so for the export market.

At the start of each agricultural year in November the hydraulic committee decides on the total area that can be safely irrigated in the district and by each of the eleven modules. This total area is derived from the combined volume of water in the four storage dams. The committee next sets the number of times that irrigation services can be delivered to each farmer in each season, typically 3-5 times in the winter and once in the summer. Each module receives its allocated share of surface water over the year and, given the module's ability to restrict the area irrigated by users, farmers can request water at any time within the constraints of the seasonal maximum. Farmers pay a fee to their water user association prior to each individual irrigation and get a receipt that has to be shown to the channel-keeper before water is allocated to their fields.

Kloezen and his colleagues show a keen awareness of the political and macroeconomic context of irrigation management transfer as well as a familiarity, for example, with Fox's sociological work on building social capital from below. However, the language of common property resources, freeriding and transaction costs is absent from their reports. What we do learn is that the number of staff engaged in governance, OM&M, administration, monitoring and evaluation *rose* by 13 per cent between the 'before' and 'after' transfer periods of 1992 and 1996. The proportion of persons employed by the National Water Commission dropped from 100 per cent to 38 per cent. The associations feel that the Commission staff numbers are excessive and that:

...an unspecified percentage of these...staff are residual personnel that for political and labour-union-related reasons remain within the agency with no specific task. This has been one of the major reasons why the modules wanted to create the [District Federation of the user associations] which will take over management of the main system. They feel that this will be much more cost-effective than the current arrangement.

As already indicated, the water user associations hire all their own staff. In doing so they have shown reluctance to employ ex-Commission personnel. The associations argue that Commission channel-keepers and other staff were often out of control, poor workers, unaccountable, given to rent-seeking behaviour and were likely to involve the trade unions in module management in a divisive manner.

*To what degree do farmers accept that the shared use of the resource is now equitable?* The importance of the new, district-wide, hydraulic committees should be noted, but the allocation planning method has not changed at this level. However, there is evidence that correspondence between the volume of water assigned to modules and the actual volumes received by them is better than before transfer. This is probably because every association was represented on the hydraulic committee, with each module insisting on receiving the volume it had been assigned and had paid for.

With respect to individual farmers, a detailed, post-transfer study shows that in the sample area 'there is no clear bias towards head- or tail-end farmers and that all farmers receive sufficient water to meet crop requirements'. In a sample of farmers, 34 per cent said water distribution amongst them was poor before transfer and good after it, compared with only 15 per cent who held the reverse opinion.

Probably the most important contribution to equitable distribution of the resource has come from the associations' control over channel-keepers. Bribes exacted by Commission employees under the previous regime were impossible for farmers to eradicate. The downward shift in power to water users now gives them the control over rent-seeking behaviour that they lacked.

*To what degree has infrastructural operation and maintenance been improved?* With respect to operation, the post-transfer study indicated an improvement in water adequacy at field level, in the timeliness of water delivery and in access to the channel-keepers. In the sample of farmers, 40 per cent said the service provided by the channel-keepers was poor before transfer and good after it, compared with only 14 per cent who held the reverse opinion. Eighty-three per cent of farmers believe that the water user associations should now retain responsibility for operation of the main and secondary systems.

The transfer process included free concessionary use of Commission machinery such as draglines and hydraulic excavators. This was a flying start for users. Significantly, some of this equipment was in serious disrepair so the modules bought twenty-nine pieces of new heavy machinery from their user fee income and also received equipment from a World Bank programme.

Maintenance expenditure in constant peso prices almost doubled in the before- and after-transfer comparison. There was an approximately threefold increase in the desilting of secondary canals and drains. Farmers' perceptions of the condition of the irrigation and drainage network were that the network had improved considerably.

*To what degree have negative externalities arising from individual farmers' actions declined since the user associations were set up?* Kloezen and his colleagues address this question only with respect to overabstraction of groundwater. By customary practice, modules exclude from the surface-water supply those areas that have access to wells. But groundwater users are permitted to use the canal network to distribute their pumped supplies.

Unfortunately the aquifers in the district are definitely being overexploited by about 20 per cent of annual recharge. By 1996 the water table was falling by 2-5 m annually. Groundwater mining does not appear to have become worse since transfer, but transfer has been unable to diminish it.

To conclude this case study, the new arrangements for the financing of capital and current costs are examined. We have already seen that the association pays fees to the Commission for its services, that the association charges the farmer for each irrigation, that there has been a decline in bribes paid to channel-keepers and that charges to farmers have been used not only for prime costs but also for equipment purchase.

Forty per cent of farmers surveyed report that payment procedures are now less cumbersome compared with only 2 per cent holding the reverse opinion. Sixty-nine per cent stated that bribery of channel-keepers had been reduced. But many farmers still come to the module office to complain of keepers demanding private payments and this has often led to workers being dismissed. Keepers are also rotated between module sections to prevent the growth of patron-client relations. Nevertheless as many as 30 per cent of farmers surveyed were willing to admit that they bribe keepers to irrigate more than the farmer's entitled area or to get water at different times from those programmed.

So the indirect charges are higher than the official ones-as are the wages of keepers. Under the Commission, charges were way below prime costs, let alone the full cost of the service. To prepare the way for a more self-reliant organization, the Commission increased the charge by some 400 per cent 2 years prior to transfer. It then remained unchanged in nominal pesos through and after transfer. But inflation halved the real value of the charge in the first 2 years of the associations' existence.

What is undoubtedly impressive is that in terms of the ratio of fees actually collected to actual OM&M expenditure, both given in pesos at current prices, the rate jumped from an average of 50 per cent in 1989-91 to 123 per cent in 1993-6. These facts remind us that in irrigation economics the estimation of the cost of water to farming families and agribusinesses should be based on the payments *actually* made, not on those that are *supposed* to be made. Kloezen and his colleagues never attempt to measure what proportion of the full cost of irrigation services is met by farmers' actual payments, but it seems likely to be considerably less than 100 per cent. This raises the question: 'What organisation in future will be responsible for infrastructural investment and rehabilitation, how will it raise the required finance and how will these costs be repaid?'

For the 1996-7 winter season the associations used the hydraulic committee to raise the irrigation fee by more than one-third. They also pushed the Commission to transfer OM&M responsibility for the main canal to a newly created District Federation of the user associations. In 1997 the average percentage of farmers' irrigation fees paid to the Commission was correspondingly slashed from 25 per cent to 9.5 per cent.

**Tradable Abstraction Rights in Australia**

Up to this point, consideration of abstraction has covered the right to abstract and the charges that a catchment agency might levy on farmers for the enjoyment of that right. *Tradeable* abstraction rights (TARs) have been discussed only briefly-although long enough to recommend that, with appropriate safeguards, such rights should be widely introduced. This case study reviews the introduction since the early 1980s of TARs within irrigated agriculture by the State of Victoria in Australia. The case study updates by 4 years onc that I published in 1997.

In every catchment, some users of water will value their abstraction rights more than other users, for example because they employ water more productively. In that situation, trading in such rights can be financially attractive to both the existing owner of such rights and the party that wishes to see them transferred. To understand trading in abstraction rights we need to come to grips with the relevant law, with differential valuations of water and with the legal, hydrological and engineering means for implementing transfers. In this field, valuable work has been carried out by Robert Hearne and William Easter, in Chile, although when they were writing their report less than 1 per cent of all abstractions operated through such water markets there.

In Australia, the practice is now extensively developed and there the main objective of TARs has been the more efficient use of scarce irrigation water. The Australian water economy is said to be in its mature phase, where the long-term average total cost function is rising sharply, where there is intense competition for existing supplies and where the hydrosocial infrastructure requires costly rehabilitation.

Victoria is one of the constituent states of federal Australia and lies in its south-east corner. The dominant physiographic feature is the Great Dividing Range, running east-west in the eastern two-thirds of the state. This creates a rainshadow in northern Victoria, where average annual rainfall is 450 mm in the Goulburn-Murray Irrigation District (GMID). This is the largest irrigation area in Australia, covering some 820,000 ha and it represents the bulk of water trading in the state. Irrigation is sourced from reservoirs on regulated surfacewater flows. Meat and dairy products are the principal agricultural output; higher-valued crops such as horticultural products are constrained by the red-brown soil type. Surface-water resources have been extensively developed in the past and any further supply growth would require inter-basin transfers. The GMID is principally designed to provide security of supply during a prolonged series of drought years. The main water dams are operated on a carry-over basis, where water is accumulated in years of high river flows for use in drier years.

Australia's modern history of property rights in water begins in the 1880s, when Victoria introduced new water laws based on the recommendations of a Royal Commission chaired by Alfred Deakin. He proposed that water allocations should be tied to the land, that rights to water should be vested in the British Crown and that allocations to landholders should be the responsibility of the state governments. Riparian owners retained limited common-law rights for domestic use, stock watering, gardens and a maximum of 2 ha of irrigated land for fodder crops. Over the next 15 years, similar legislation was introduced in the other federal states. Using the new legislation, the states ventured into large irrigation projects, providing water to farmers far below average total cost. Areas were established in which government constructed massive irrigation and drainage infrastructures, such as the Goulburn-Murray Irrigation District itself, the Murrumbidgee Irrigation Area in New South Wales and Riverland in South Australia.

The subsidy of water led to overuse and this was consolidated by the historical overallocation of water to irrigated farms. As argued in the work of Pigram and co-workers: This has its roots in the social objectives of past governments for the development and use of water resources. The overriding objectives were an equitable distribution of water among farms and the promotion of regional development and closer settlement of inland areas. All farmers were considered to have an equal right to the available water, irrespective of how much was needed to irrigate the proposed crops to be grown on the farms, or of any consideration of how efficiently the water would be used. As water rights became capitalized into land values, individuals had the economic incentive to retain their entire water rights through demonstrating a history of use. In many cases, this actually translated into a history of over-use.

By 1990, the arrangements in the GMID were that irrigators paid for a water right based on both the amount of land they held that was suitable for gravity-fed irrigation and the irrigation district in which they were located. An annual charge applied to the water right, invariant with the volume of use.

Subject to availability, 'sales water' was also available, at a volumetric charge based on the water right charge. These water rights in their specific form were assigned by a licensing system for a fixed period of 15 years, with an expectation of reissue. The authorities could vary the volume licensed, usually in times of shortage. However, before 1987-8, no arrangements existed for *transferring* farmers' abstraction rights.

An Australian interest in TARs (also known as tradeable water rights or transferable water entitlements) began to surface in the mid-1970s. For some, the introduction of TARs seemed to offer a new flexibility to existing arrangements, with clear economic and environmental advantages. *First,* we have already seen that the maturing of the Australian water economy in recent decades is associated with high average total cost for the long-term supply curve. So, TARs offered a new direction that, by reallocating water supply, would reduce the pressure for aggregate supply expansion. This reallocative effect would not be merely within the farming sector. There was also the prospect of reducing agricultural overuse to open up supplies for a range of urban and industrial uses.

*Second,* supply reallocation within agriculture was a major objective. There was a widespread belief by the mid-1980s that TARs would switch water from lower to higher water-productivity uses in the farming sector. In Australia, at that time, the agricultural sector accounted for 80 per cent of total water use. In Victoria, each year, up to one-third of irrigators were using less than their full water right allotment. Specific switches into river red gum watering, salinity dilution and dairy farming were forecast. *Third,* environmental benefits were expected from transferability. The new policy promised to reduce the scale of infrastructure construction for inter-basin transfers, with all their negative externalities. Development proposals were meeting increased resistance from environmental groups. Moreover, reductions in agricultural overuse promised to have a positive environmental effect through a reduction in waterlogging, salinization and biocide dispersion.

The redistribution of abstraction rights could have been sought by the administrative processes of the licensing system. But this would have been met by political resistance from the farms on which volumes were to be reduced and land values consequentially cut. Tradeable abstraction rights offered these farms a pay-off. For landholders wishing to move out of irrigation, but to remain in dryland agriculture, transferability allows water entitlements to be sold separately from the land. Previously, the options for such irrigators were either to cancel their water rights licence (or not renew it) and get nothing for the right, or sell the entire irrigation holding and buy a dryland property elsewhere. Transferability can permit a more flexible retirement plan for an irrigator, or facilitate a long-term change in enterprise or financial structure of the farm business.

*The fundamental requirement of a workable and efficient market in abstraction rights is a clear specification of property rights in water such that*:

- Rights in land and rights in water must be separable;
- The volume of water that an individual or institution has available for transfer must be clearly stated, as well as any special conditions on its use-such volumes are likely to be conditional on rainfall, or surface flows or groundwater stocks;
- The right to transfer such water at a privately negotiated price must exist;
- The period over which such a transfer is deemed to be effective, temporary or permanent, must be known;
- The power of government to restrict or terminate abstraction rights at a future date must be known.

In Victoria, TARs were cautiously introduced in 1987-8 with a *temporary* scheme, before permanent transfers were considered. In the GMID the transfer period was to be for 1 year, only between irrigators and only within the same supply system. There was no volumetric limitation but stock and domestic allocations had to be retained. A state agency assessed possible third-party effects and could refuse the transfer if these were significant. The arrangements were not to affect significantly delivery and drainage channel capacity or salinity. Finally, the price was determined between buyer and seller but the agency's fixed fee was A$70.

A new Water Act in 1989 permitted *permanent* transfers of abstraction rights between farmers, with effect from 1991-2. Transfers out of agriculture were still proscribed. However, despite these references to 'permanence' in trading, the state assignation of water rights through the licensing system, meant that a purchaser of abstraction rights could not be assured in law that such water rights would continue indefinitely. The state government retained long-term flexibility in its management powers.

TARs impose certain transaction costs. In the GMID case they seem to be modest because of the legitimacy of the institutions, the support from farmers for the general approach to the new rights and the competence and honesty of Victoria's state administration. Incremental supply costs also exist if distribution and drainage structures cannot handle the increased water volumes. However, in Australia, water agencies have taken the easy administrative option of simply refusing transfers where existing capacities would be exceeded.

The outcomes of Victorian TARs in the early years after they were first introduced are considered next. The scale of change can be measured by the volume of abstraction transfers as a percentage of all abstractions. With respect to benefits, we are looking for: the avoidance of new infrastructure expenditure as a result of the redistribution of abstractions within and between user categories; an increase in agricultural productivity as a result of abstraction redistribution within farming; and a fall in waterlogging, salinity and pollution from run-off because of the reduction in agricultural overuse.

In the 2 years 1987-8 and 1988-9 the trading of abstraction rights in Victoria (most of it in the GMID) averaged 25 mcm, less than 2 per cent of total abstractions. Price per megalitre ranged from A$8 to A$20. Gross margins increased. But at this point the judgement was that 'Despite widespread endorsement of the concept in Victoria, transfer activity has been sporadic in that state'.

In Victoria by 1990-1 trade was still 'insignificant compared to the total volume'. In a mail survey of all 299 permanent water transfers within the GMID up to 1994, with a 63 per cent response rate from buyers, purchases made up between 0 per cent and 2.3 per cent of total allocations when the responses were grouped by district. What the survey showed was that sellers were releasing sleeper water (*i.e.* unused allocations) and that farmers in financial distress were also selling. In a smaller number of cases, farmers either wished to cut their irrigation agriculture, or to farm without irrigation, or to retire. In respect of purchasers, the single most important reason for buying water, applicable to 65 per cent of respondents, was that they 'wanted to secure existing crops against future drought'.

Stringer suggested that in Victoria as a whole in 1991-4, the average percentage of allocated water traded in each of those years was 0.33 per cent. In the Murray region as a whole in 1993-4, the volume of transfers was 17 mcm for temporary transfers and 8 mcm for permanent transfers. For 1994-5, these figures changed dramatically.

Temporary transfers soared to 265 mcm whereas permanent transfers fell to 2 mcm. At this point it could be argued that by the mid-1990s TARs in Victoria had created a space for single-season switching of abstraction, most active in years of low rainfall. By the year 2000 Australian statistical reporting of TARs had become much fuller and more informative as a result of work by the Australian National Commission for Irrigation and Drainage. Its *Benchmarking Report* contains data for nine Goulburn-Murray systems known as Murray Valley, Shepparton, Central Goulburn, Rochester, Pyramid-Boort, Torrumbarry, Nyah, Tresco and Woorinen. In all of these except the last the percentage of surface-water irrigation supply points that are metered lies in the range 87-98 per cent.

A broking service exists in all nine systems to facilitate the trading process between users. In 1998-9 temporary trading took place in all nine systems, averaging 11 per cent of total abstraction entitlements for that year. Six of the nine systems reported permanent trading that averaged 2 per cent of entitlements in those six areas.

From the perspective of this book, the material on why one set of irrigators sell their TARs and others purchase them is of particular interest. A substantial paper by Björnlund and McKay throws some light here. The authors state that the main use of irrigation water is in the cultivation of pastures for dairy cattle. Some activity is found too in horticulture and viticulture. Mixed grazing and cropping is said to have low-value water use.

Purchasers of tradeable abstraction rights were shown to be seeking to ease their reliance on annual sales water and temporary water market purchases. TAR sellers are smaller than buyers in terms of irrigated area; water trading can be seen to be integrated with a long-term consolidation and amalgamation of farm properties.

Sellers were asked how they used the cash proceeds of their permanent sales-note that a permanent sale does not imply the sale of a farm's entire water entitlement:10 per cent was invested in laser grading, 5 per cent in improved drainage and another 5 per cent in other irrigation equipment. Fifty per cent of the sample said the money was used for general consumption purposes, related or not to the farm as an enterprise; 21 per cent went to debt reduction; 6 per cent to non-irrigation equipment; and 3 per cent for the purchase of another farm.

Clearly the subject is ideally suited to an irrigation economics case study, for both the permanent and temporary markets, into the differential motivation of buyer-seller *pairs*.

## A BLOCKED CHANNEL

One aspect of this blocked channel concerns payments for abstraction rights. I have given a very full account of the criteria that can be used by a state authority in constructing an abstraction tariff for farmers and the relevance of such a tariff to effective catchment management. But there is no accompanying case study to test these ideas-merely brief, illustrative examples. In contrast there is an extended discussion of the recent history of tradeable abstraction rights in Victoria, but the theoretical material is adhoc and fragmented.

Despite this the research reports of Wim Kloezen, Carlos Garcés Restrepo and Sam Johnson seem to me to be wholly successful in coming to grips with the dynamic institutional changes of this catchment. Perhaps it is because they were a team with varied disciplinary backgrounds, which is so important in water resource planning.

Perhaps it is that irrigation management transfer is quintessentially a compound of private economic interests and political power set within a given hydrological context-and that Kloezen and his colleagues have a good grasp of hydropolitics. In contrast, the language of many economists is so distant from the complex drives of individuals, families and organizations and so wedded to a vision of the determining interplay of measurable quantities, that they fail to understand the nature and exercise of power.

## ABSTRACTION CHARGES IN RIVERS AND LAKES

Rivers, lakes and aquifers are common property resources, except where access to them is privatized. This is why common property resource analysis, including the appraisal of transaction costs, is of such great relevance to a full understanding of the management of water resources. In particular, rivers, lakes and aquifers are resources from which water is abstracted for use in agriculture, industry and households. They are also resources used as sinks for the disposal of waste water and drainage water.

Abstraction usually takes place where the abstracting actor enjoys formal legal rights to do so, or where abstraction is an accepted customary practice. In such cases the abstractor may be required to pay a charge either to a public agency or to a private owner of rights in water. Abstraction rights may also be traded. One objective of these two sections is to provide a general economic analysis of the subject capable of international application to specific case studies. The second objective is to recommend the most appropriate basis for government abstraction tariffs. The scope of abstraction charging includes water consumers abstracting directly for their own use, such as farmers, mining companies and manufacturers. One also includes water service companies that abstract not for their own use but in order to provide a public water supply to irrigators and to urban areas.

A useful analytic starting point is to establish that abstraction charges are a form of economic rent. The modern theory of rent was first developed in the early nineteenth century. Its application was to British agriculture and the common situation where capitalist farmers used land owned by the aristocracy and the gentry and paid rent for that right. At that time, agriculture contributed about one-third of Britain's total output. The classical school of economists argued that the appropriation of land has the consequent effect of the creation of rent. Thus, Ricardo writes: 'Rent is that portion of the produce of the earth, which is paid to the landlord for the use of the original and indestructible powers of the soil'.

It is true that we may now doubt that any power of the soil is indestructible. Nevertheless, since Ricardo's time, economists have used rent theory whenever they are dealing with a natural resource of economic value that can be privately appropriated and that is in restricted supply. Clearly, this makes rent theory applicable to natural supplies of freshwater. These flows are of economic value, rights in their abstraction can be privately appropriated and groundwater as well as surface water are in restricted supply in any given year.

However, the supposed inelasticity of the water supply deserves closer scrutiny. Within a catchment, short-term elasticity is perversely high where substantial stocks are held in a catchment's reservoirs or where aquifer stocks are high and groundwater pumping capacity is not fully utilized. But it is the elasticity of the planning supply function of water over the medium and long term that is at issue here. It is certainly true that total rainfall in the catchment less evaporation sets a hydrological constraint on long-term catchment abstraction, as Dubourg suggests. But the recycling and reuse of water weakens this natural barrier whilst the desalination of sea water and abstracted water imports from another catchment, can smash it. The problem is that the wider one casts the net for additional supply, the greater are its unit costs. It is the exponential character of this planning function which maintains the truth of the statement that, in the majority of the world's catchments with substantial populations, abstracted water is indeed in restricted supply.

Returning to the classical theory of differential rent, this assumed that competition existed between farmers for access to land and between landowners in supplying land. In such a situation, it is theorized that land rent on the least fertile tract of land worth cultivating would be zero and here the farmer would earn the going rate of profit on capital for the private sectors of the economy as a whole.

Land which was more fertile (or better located) would bring the owner a higher flow of rental receipts such that the farmer, after paying this higher rent, would still receive only the going rate of profit. Rent could therefore be seen as the transformation of the surplus profits of capital, driven by the competition for land of differential fertility.

Ricardo had only a limited interest in non-competitive markets in the supply of land for rent. As a consequence, his theory is essentially demand-driven. This neglect of the supply-side determinants of land rent restricts the scope of differential rent theory in its application to abstraction charges levied by government.

Such charges typically exist where an agency of the state licenses abstraction to specified individuals and institutions and where no property right exists to abstract ground and surface water without such a licence. Here, property rights on the supply side are assigned by a state monopoly. So analysis needs a means to interpret the legislative practice of charge-setting by state institutions in different catchments, regions or countries.

Having argued that abstraction charges are a form of economic rent, but that differential rent theory does not provide a basis to explain charge-setting by the state, the best way to proceed is to develop a taxonomy of charge-setting principles. Such principles can then provide the basis for interpretive empirical work in specific catchments, as well as a starting point for policy review of existing legislative practice.

Various charge-setting principles are set out below and each is glossed in turn. Note that these charging principles are in respect of the right to appropriate the scarce resource for outstream purposes. They do not include the cost of supply-side infrastructure.

*Nor are the public-sector costs of flood management, navigation, etc. included here*:

- *No charge:* This is the lower limiting case. In the majority of the world's countries, government abstraction charges simply do not exist. In some cases the argument is that, since surface water and groundwater are a gift of Nature or of God, the state has no business in taxing it. This may explain the situation in Scotland, for example. In other cases, what is lacking is the administrative capacity to levy the tax.
- *A revenue-maximizing charge:* This is the upper limiting case. In principle government could raise all its revenue requirements from this single tax. In practice, of course, such a principle is never applied.
- *A market-clearing charge:* In countries and regions that are arid or semi-arid, or where levels of water consumption are high compared with effective rainfall, users may wish to gain access to more water than is available, at least in the absence of inter-catchment transfers. In this case, government may put in place a demand-management policy in which a general abstraction charge is applied that, although it is not revenue-maximizing, does broadly match the demand by abstractors to the annual flows available. No pure examples exist of this, the closest being the public auction of tickets for fixed time and flow in the centuries-old water market of Alicante in Spain. Schiffler and his colleagues at the German Development Institute recorded for the case of Jordan that supply-fix policies in the early 1990s were under considerable pressure and that a demand-management philosophy was taking shape. Politically, it was impossible to tax water abstracted for use by farmers, but an abstraction tax on industry had been introduced there.
- *An environmental regulation charge:* In this case one is considering a country which has a well-developed national policy for water resource management. The necessity for a regulator of the freshwater environment is accepted and abstraction charges are levied and hypothecated to finance the costs of regulation. ('Hypothecation' is a term widely used in economics to refer to cases where government income from a defined tax is reserved for a specific expenditure category.) This charge may take the form of an average total cost levy equal to the total financial costs of regulation (including compensation payments for rescinded abstraction rights and any other miscellaneous items) divided by the total volume of water abstracted during the year.
- *A Pigovian charge:* Here, the principle is that government should set a charge which is differentiated according to the external costs

imposed on society by each class of abstractor. K. William Kapp later formulated the parallel concept of 'social cost', covering 'all direct and indirect losses sustained by third persons or the general public as a result of unrestrained economic activities'. An example of the kinds of damage done by overabstraction is that of the Hueco Bolsón aquifer on the Mexican-USA border. Over a period of 70 years the water table fell by 25 metres (m), resulting in increased pumping costs, subsidence and contamination by increased flows of saline and polluted waters into the freshwater source. As a general rule, whilst abstraction charges may include a component in recognition of overabstraction, the *differentiation* of the charge on the basis of the monetary evaluation of the external costs never takes place because of the extreme difficulty of measuring them. Indeed, many institutional economists even deny that such measurement has any meaning. Kraemer refers to the overwhelming problems of applying a Pigovian tax in his introduction to the development of German abstraction charging after 1988.

- *An incentives charge:* Incentive charging in a catchment or region can be defined as the use of a water tariff-a table of fixed charges-to give price signals to abstractors that reinforce water resource management based on environmental standards and regulatory controls. The best-developed system of incentive charging is probably that of the Federal Republic of Germany. An incentive tariff can include the following components:
  - A licence fee to have a new water abstraction installation approved by the regulatory authority. For example, Foster in their comprehensive introduction to the hydrogeological, legal and administrative aspects of groundwater licensing in Latin America indicate that installation fees are common and are hypothecated.
  - A charge per unit volume. This may be invariable with total volume consumed. Alternatively, it may rise with the volume drawn off. The total charge payable may be based either on the licensed volume or the actual volume abstracted. As Schiffler point out, a significant drawback of a rising block tariff is that it hits hardest those abstractors requiring large volumes simply because of the size of the farm or factory. It should also be observed that abstraction charging is never imposed where the installation's capacity falls below a minimum level set by government. This is the case at Keveral Farm.
  - A charge reduction for the quantity of water directly recycled to surface waters after use. To take the example of Didcot power station in the UK, in the mid-1990s this had a licence to abstract

142 million litres of water per day. The licence required 50-66 per cent of the water abstracted to be returned to the river, depending on flow conditions. Unit price was lower because of this non-consumptive use. In contrast, spray irrigation provided virtually no return flow to river or aquifer and so was undiscounted.

- A charge that is greater for higher quality water. This is found in Germany for water drawn from deep aquifers.
- A charge which varies with the seasons. The volumetric rate is higher in those months when demand is greater and higher too when precipitation is less.
- A charge which is greater for certain locations. These include
  (a) Upstream sources, because the length of the river exposed to abstraction impacts is greater,
  (b) Rivers, lakes and aquifers most threatened by past or present overdraft;
  (c) Regions with lower effective rainfall.

## INTENSIVE IRRIGATION AND WATER LOGGING

Most anti-dam movements in india have emerged from move ments of people facing displacement due to the submergence of large areas upstream of dam sites. These struggles are expressions of a conflict of interests between those who bear the social and ecological costs of dams and those who benefit from them.

However, large dams have diverse and complex ecological impacts. and they often generate environmental costs for those very groups who are supposed to be the beneficiaries. Waterlogging and salini sation are twin problems caused by the wasteful use of water.

Waterlogging is caused by the interaction of a large number of factors such as irrigation intensity, soil characteristics, drainage. seepage from reservoirs, distributaries and field channels. Since large-scale irrigation systems are linked to the uniformity of water distribution, which enforces the uniformity of cropping patterns. and uniformity in the landscape, waterlogging becomes inevitable in areas with undulating topography and water retentive soils. In such cases, farmers who are supposed to be the 'beneficiaries' become victims of irrigation projects and irrigation authorities. In areas where irrigation has led to the transformation of productive lands into waterlogged wastelands, conflicts arise between farmers and the state.

The 'Mitt) Bachao Andolan' in the Tawacomman area is an example of such conflicts." In the Krishna basin, conflic generated by irrigation projects were highlighted by the farmer agitations in the command area of the Malaprabha project.

The Malaprabha project was completed in 1972-73. With the introduction of canal irrigation, nearly 2,364 hectares of land i the project area has become waterlogged and saline' Before the introduction of perennial irrigation, the undulating semi-arid land in the project area was used for growing water prudent crops like jowar and pulses. Due to a sudden change from rainfed agriculture to intensive canal irrigation, the low lying areas have become waterlogged. The cultivation of water demanding crops like hybrid cotton has aggravated the problem. In addition, seepage from canals has also raised the water level. The Malaprabha project includes a storage dam of 1,068 million cum capacity near Saundatti in Belgaum district which feeds the 138 km Malaprabha right bank, 168 km left bank canal, and the Kolachi right bank canal. The soils in the command area are black cotton soils, which have high water retention capacity and are prone to waterloggin, Intensive irrigation of black cotton soils has been known to be prescription for creating wastelands. While irrigation has bet viewed as a means to improve land productivity, in cases like if Malaprabha command area, it has led to the destruction of productivity.

Further, the shift from rainfed food crops to an irrigated cash crop like cotton was expected to improve the prosperity of farmer However, it led to indebtedness as well as loss of fertile fan. through waterlogging.

To utilise the irrigation waters, farmers began to cultivate 'Varalaxmi' cotton which was initially sold at ₹ 1,000 per quintal. Farmers took loans from banks to develop land, purchase seeds, chemical fertilizers and pesticides. The total loan taken by the farmers increased from ₹ 50 lakhs in 1974 to over ₹ 5.5 crores by 1980. The prices of chemical fertilizers increased from ₹ 75 to ₹ 103 per bag. The cost of the Varalaxmi seed increased from ₹ 60 to ₹ 170 per kg. In the meantime the price of cotton crashed from ₹ 1,000 per quintal in 1974 to ₹ 350 per quintal in 1980.

While farmers were caught in the trap of unfulfilled commercial promise, banks demanded repayments of loans, and the irrigation authorities demanded a development tax known as betterment levy of ₹ 500 to ₹ 1,600 per acre. The water tax was raised from ₹ 18 to ₹ 30 per acre for jowar, ₹ 18 to ₹ 50 per acre for Varalaxmi. A tax of ₹ 10 per acre was fixed even if water was not utilised. For the farmers, this amounted to gross injustice, since they had not benefited from the irrigation project. In addition, the compensation for acquiring land for the dam and canals had not been paid to 75 per cent of the farmers even after seven to eight years.

The farmers therefore organised themselves as the Malaprabha Niravari Pradesh Ryota Samvya Samithi' (Co-ordination Committee of Farmers of Malaprabha Ittihsyrf Area) in March 1980. When the local authorities did not pay heed to the farmers' demands, they launched a non-cooperation movement for nonpayment of taxes. The authorities responded by refusing to issue the certificates required by the farmers' children in order to in schools and colleges.

On 19 June, the farmers went on a hunger strike in front of the Tebsildar's office in Naragund town. On 30 June, 10,()00 farmers collected to support those on hunger strike. On 7 July, a massive rally was organised in Navalgund, and the farmers went on a hunger strike. Seeing that no response was forthcoming from the authorities, the farmers organised a 'bunch' on 21 July. When 5,000 to 6,000 farmers had gathered in Navalgund, their tractors were damaged and the rally was stoned. The protest then took a violent turn. The angry farmers seized the irrigation office department, burnt down one truck and fifteen jeeps. The police in turn opened fire and a young boy, Basappa Shivappa of Algavadi, was killed on the spot.

In Naragund town, the police opened fire at a procession of 10,000 people, shooting one youth. The protesting farmers responded by beating a police officer and a constable to death. The protests rapidly spread to Ghataprabha, Tungabhadra, and other parts of Karnataka. During the protests thousands of farmers were arrested and forty were killed. Finally, the government had to put a moratorium on the collection of water taxes.and the betterment levy. According to a rough estimate, the concessions granted to farmers to end the Malaprabha agitation amounted to ₹ 85 crores.

However, the high costs of irrigation in the Malaprabha project have been forgotten. No lessons have been drawn for planning water projects, and bureaucrats, technocrats and politicians continue-to get carried away by the euphoria for large dams and intensive irrigation projects. The creation of waterlogged wasteland through intensive irrigation is not specific to the Malaprabha command area. Compared to other projects in the Krishna basin, waterlogging, salinity and alkalinity are most serious in the Tungabhadra project. Nearly 1,500 hectares of land is likely to become waterlogged under the left bank canal.

In the right bank canal 6,000 hectares have been affected by waterlogging. Under the right bank high level canal 12,000 hectares have been affected by waterlogging. In all, 19,500 hectares have been destroyed by waterlogging in the Tungabhadra project within an irrigation period of thirty-five years.

In the Bhadra project, of 1,24,392 hectares irrigated, 7,900 hectares have been devastated by waterlogging. In the Malaprabha project, of the total potential of 2,12,086 hectares, only 12,186 hectares have been actually irrigated. Of this irrigated area, 50 per cent has been waterlogged.

In the Ghataprabha project, where a higher average has been irrigated than what was actually planned, out of 3,58,542 hectares irrigated, 19,948 hectares have been devastated by waterlogging. In Andhra Pradesh, under the Nagarjunasagar project (NSP) the groundwater level has risen alarmingly within ten years, thereby indicating a trend towards waterlogging. In Maharashtra, the Maharashtra Irrigation Commission claims that 28,000 hectares of land have been affected by waterlogging in the Deccan Canals, *i.e.*, Nira and Mutha Canals.

The irrigation commission of 1976 had estimated the total waterlogged area in the basin at 7,828 hectares (6,583 hectares in Karnataka and 1,245 hectares in Maharashtra) and 15,502 hectares affected by salinity. At present 4,45,985 hectares have been affected by salinity. Waterlogging is ecologically linked to large dams because large darns involve the transport of huge quantities of water for intensive irrigation. In fact, the primary rationale given in defence of large dams is to induce a shift from protective irrigation, which is ensured by indigenous irrigation systems, to intensive irrigation for commercial crops. The inevitable ecological impact of overuse of water for irrigation is a build up of water beyond the drainage capacity of the ecosystem. The need for artificial drainage systems arises because the natural drainage processes of the local ecosystem are violated. Waterlogging is thus a symptom of the conflict between water use in the commerciaVmarket economy, and water use for the maintenance of the water cycle including a balance between water entering an ecosystem and water leaving it. By violating the ecological laws of water flow, large dams lead to ecological destruction on the one hand and political conflict on the other.

## RIVER DIVERSIONS AND REGIONAL CONFLICTS OVER WATER

Large dams are constructed for allowing major diversions of water from the natural drainage flow of the river. These diversions result in a major change in the distribution patterns of water in a basin, especially when they involve inter-basin transfers. They therefore generate new conflicts over the distribution of water between different regions. Regional conflicts become inter-state conflicts, and are rapidly enmeshed in inter-state and centre-state politics. The Telugu Ganga Canal, which takes off from the Srisailam Dam, is probably the most conflict-ridden river diversion project in contemporary India.

Krishna is the second largest river of peninsular India. Its catchment lies in the Western Ghats and it flows east through the states of Maharashtra, Karnataka and Andhra Pradesh. Krishna is an inter-state river, and conflicts have arisen between the co-riparian states over the allocation of its waters to their respective territories for purposes of development.

The Krishna basin like other regions of India had indigenous irrigation works such as tanks, wells and anicuts. There were nearly 27,000 small tanks and diversions on the Krishna river system, mostly in Andhra Pradesh and Karnataka. To these were added new canals during the colonial period for commercial agriculture. These were:

1. The Krishna Delta Canals built in 1855.
2. The Nira Canals in Maharashtra constructed in I885 irrigating about 150,000 acres.
3. The Kurnool Cuddapah Canal in Andhra Pradesh built in 1886, irrigating 100,000 acres.

The majority of the area irrigated by the Krishna Delta Canals and Kurnool Cuddapah Canal (1.11 million acres) lies outside the basin of the Krishna. In 1951, the status of the diversion of Krishna waters was as follows: 411.4 TMCF of water was diverted annually for the irrigation of 2,302,377 acres. Of this 290.1 TMCF was used by Andhra Pradesh, 430 TMCF by Maharashtra, and 78.3 TMCF by Karnataka.

After independence, the large-scale diversion of river waters increased. In July 1951 the Planning Commission convened an inter-state conference to discuss the utilisation of Krishna waters. The dependable annual flow in the Krishna basin based on the recorded gaugings at Vijayawada was agreed at 1,715 TMCF so the balance of flow for new projects remained 970.5 TMCF which was rounded off to 1,000 TMCF and allocations were made between the different states as follows:

For the balance flow in excess of 1,000 TMCF, if any, the allocation for the above states was in the ratio 30:30:1:39. The state of Bombay was allowed to divert the waters to the west across the Western Ghats for the hydro-electric project at Koyna up to a limit of 67.5 TMCF. The agreement provided for a review of the allocations after twenty years. In 1953, states were; reorganised on a linguistic basis, Madras was divided into Andhra and Madras. In 1956 the state of Andhra Pradesh was created by the merger of parts of Hyderabad and Andhra.

As a result of territorial changes, the riparian states sharing the Krishna waters are Maharashtra, Karnataka and Andhra Pradesh. An inter-state conference was convened in New Delhi under the auspices of the Union Minister of Irrigation and Power on September 1960, to recast the allocations of Krishna waters made in 1981. However, efforts to reach an agreement among the states proved unsuccessful and widely divergent views were expressed by the different states. ~ three man commission headed by N.D. Gulhati was set up. The commission undertook the first ever attempt 'at a basin-wide survey of the technical implementation relevant to water resources development.'

As observed by Tripathi, the Commission in examining the river flow of both these rivers was greatly hampered by the lack of regular reliable and continuous observations of water discharge at various points in river... The Commission stated that the flow records prior to 1936 were based on formulae different from those followed after 1936. Therefore the Commission stated that it [was] not possible to determine the flow for 86 per cent dependability or for 75 per cent dependability or for any other criterion of dependability.

Because of the lack of adequate data of river flow, the Commission could not give positive answers to the terms of reference as regards the availability of water supplies on the river systems. State-wise allocation of Krishna waters was, therefore, not possible owing to the lack of scientifically observed data. The Irrigation Minister decided that adequate river data should be collected over a number of years and analysed continuously. However, tentative allocation was made for ongoing projects.

In spite of interim re-allocations, conflicts over Krishna waters continued with each state accusing the other of higher withdrawals from the river than its legitimate share. Maharashtra and Karnataka wanted a tribunal set up under Section 3 of the Interstate Water Disputes Act, 1956.

The Bachawat Tribunal was appointed in 1969 to resolve the Krishna water state conflicts. In 1973 the Bachawat Committee gave its award. The availability of water was assessed at 2,060 TMCF and on the basis of 75 per cent dependability, Andhra Pradesh was allocated 800 TMCF. The award fixed a formula for sharing both during surplus and lean years and was binding on the states until AD 2000.

Mrs. Gandhi the then Prime Minister consulted the co-riparian states to provide drinking water to Madras which had been facing severe shortages. The three states readily agreed to part with S TMCF each. The Chief Minister of Andhra Pradesh later hailed the Krishna water supply scheme to Madras as the Telugu Ganga. The agreement was reached on 14 April 1976. On 17 October 1977, it was agreed that a 330 km long open canal would carry 15 TMCF to Madras.

While the decision to supply drinking water to Madras was agreed by all the states, conflicts arose when Andhra Pradesh decided to use the Telugu Ganga project for irrigation. The ₹ 850 crore project now envisages extension of irrigation to 5.75 lakh acres in three districts of Rayalseema-Kurnool, Cuddapah and Chittoor and one district in the Andhra region-Nellore, in addition to the supply of 15 TMCF of drinking water to Madras. Nearly 42.4 per cent of Kurnool district lies in the Krishna basin. Cuddapah and Chittoor as well as the rest of Kurnool lie in the Pennar basin.

Karnataka had questioned the diversion of water outside the basin to the KWDT arguing that only in-basin needs should be considered in determining a state's equitable share, a state should be permitted to divert its share of water outside the basin. Andhra Pradesh maintained that out of basin needs are a relevant factor and that diversions outside the basin for irrigation needs only should be permitted. Using precedence from the American Law, the Bachawat Tribunal held that the diversion of Krishna water outside the basin was legal. The river basin as an integral unit was thus substituted by the state as an administrative unit. The conflicting demands and distributive patterns emerging from the integrity of the basin versus the integrity of the state ifs a major reason for inter-state conflicts between riparian states not getting fully resolved.

Another fundamental reason for the intractable nature of river conflicts arises from the rights established through the priority of Project use in time and the rights based on the priority of need in the long term. Andhra Pradesh contends that the diversion of an additional 275 TMCF of Krishna waters to feed the districts of Kurnool and Cuddapah in Rayalseema for irrigation of 2.75 lakh acres, is within the scope of the Bachawat award as the Tribunal permitted Andhra Pradesh to take advantage of the surplus flows down the river at Vijayawada. As the Bachawat Tribunal stated, 'the state of An&a Pradesh will be at liberty to use in any year the remaining water that may be flowing in the Krishna river'.

Karnataka has objected to the Telugu Ganga irrigation scheme on the grounds that its own projects to harness Krishna waters are still incomplete and what appears to be excess, currently, will be used in the future. Karnataka has made it clear that surplus Krishna waters would not be available for the Telugu Ganga project. Maharashtra has also opposed the Telugu Ganga project on the ground that it violates the inter-state agreement reached in October 1977. The government of Maharashtra has observed that the state has vast chronic drought affected areas. Almost 75 per cent of the Krishna basin area in Maharashtra is drought prone and the state has plans to use the Krishna water allocated to it by the Krishna Tribunal. It has, therefore, to make sure that at the time of review of the award, its legitimate claim to the surplus available water in the Krishna river is not in any way jeopardised by pre-emptive efforts to commit this surplus water to projects like the Telugu Ganga. Karnataka- and Maharashtra governments are resisting the project on the grounds that Andhra Pradesh has already used its allocation and the Telugu Ganga project would enable Andhra Pradesh to establish its right on larger volumes of water through prior utilisation.

Andhra Pradesh has already invested ₹ 200 crores and has 5,000 labourers working on the construction of the canal. Of the 406 km length of the canal, 190 km pass through the reserved forests of the Nellamali Range, for which central environmental clearance has not been obtained so far. At present the work is confined to reservoirs and canals in the non-forest areas. Water for the project is to be drawn from the Srisailam Dam through the head regulator at Pothireddypadu, which has a total carrying capacity of 11,000 cusecs.

The first 16 km of the canal is shared with the Srisailam right branch canal. The common canals run up to Bankacherla cross regulator where the Srisailam right branch canal and the Telugu Ganga Canal branch off to the right and left, respectively. The Telugu Ganga Canal is in fact the old Srisailam left branch canal extending into the Segileru Valley.

Water to be drawn for the Telugu Ganga project is to be stored in four reservoirs at Yellgodu, Brahamasagar, Somashila and Kandaleru. At Mithakanda? near the Bankacherla regulator, a 100 feet high ridge divides the Krishna and the Pennar basins, where the water would be transferred outside the Krishna basin. The canal would pass through Kurnool and Cuddapah districts from where the water would flow into the Pennar river at Chenumukapalli. The flow down the river would be picked up at the Somashila Dam and passed on to the Kandaleru reservoir before it reaches Madras. The construction work continues even though the controversy over the Telugu Ganga project remains unresolved.

Andhra Pradesh derives its legitimacy from two arguments. First, it claims it is using only surplus waters for the project, and the right to surplus waters had been granted to it by the Tribunal. Second, it claims that if there is scarcity, then the arid drought prone regions of Rayalseema should not be asked to sacrifice irrigation waters. Instead, Maharashtra should be asked to stop diverting large volumes of water out of the Krishna basin into the Arabian Ocean for power generation for industrial centres. The river Krishna emerges in the Western Ghats and flows eastward down the gentle slopes. The western face of the Western Ghats falls steeply down altitudes of 1,000 to 2,000 feet, providing excellent sites for power generation. However, the water used for hydro-electricity has to be diverted out of the basin, and dropped into the sea. Currently, the power projects in Maharashtra which divert water westwards are the Tata and Koyna Hydel Projects. The former diverts 42.6 TMC and the latter diverts 67.5 TMC.

In this conflict between the demands for power generation and the demands for irrigation in drought prone areas, the Krishna Tribunal protected existing diversions while giving priority to irrigation for future use. As it stated:

In the Krishna Basin, water is a scarce commodity. Westward diversion of water for power generation seriously restricts the use of water for downstream irrigation.... Power for Bombay and Maharashtra industry is generated at the cost of depriving the low rainfall areas on the eastern side of the water solely needed for irrigation.

The Tribunal, however, allowed the expansion of hydel projects on the condition that over a period of twenty years they would return to the existing capacity. When the next Krishna Tribunal meets in the year 2020 to review the sharing and utilisation of Krishna waters, the concepts of justice and rights as related to water will have undergone dramatic changes, as will the basin itself.

## SOURCES OF IRRIGATION WATER AND GROUNDWATER

Sources of irrigation water can be groundwater extracted from springs or by using wells, surface water withdrawn from rivers, lakes or reservoirs or non-conventional sources like treated wastewater, desalinated water or drainage water. A special form of irrigation using surface water is spate irrigation, also called floodwater harvesting.

In case of a flood (spate) water is diverted to normally dry river beds (wadis) using a network of dams, gates and channels and spread over large areas. The moisture stored in the soil will be used thereafter to grow crops. Spate irrigation areas are in particular located in semi-arid or arid, mountainous regions. While floodwater harvesting belongs to the accepted irrigation methods, rainwater harvesting is usually not considered as a form of irrigation. Rainwater harvesting is the collection of run-off water from roofs or unused land and the concentration of this. Some of Ancient India's water systems were pulled by oxen.

Around 90 per cent of wastewater produced globally remains untreated, causing widespread water pollution, especially in low-income countries. Increasingly, agriculture is using untreated wastewater as a source of irrigation water. Cities provide lucrative markets for fresh produce, so are attractive to farmers.

However, because agriculture has to compete for increasingly scarce water resources with industry and municipal users, there is often no alternative for farmers but to use water polluted with urban waste, including sewage, directly to water their crops. There can be significant health hazards related to using water loaded with pathogens in this way, especially if people eat raw vegetables that have been irrigated with the polluted water. The International Water Management Institute has worked in India, Pakistan, Vietnam, Ghana, Ethiopia, Mexico and other countries on various projects aimed at assessing and reducing risks of wastewater irrigation. They advocate a 'multiple-barrier' approach to wastewater use, where farmers are encouraged to adopt various risk-reducing behaviours. These include ceasing irrigation a few days before harvesting to allow pathogens to die off in the sunlight, applying water carefully so it does not contaminate leaves likely to be eaten raw, cleaning vegetables with disinfectant or allowing fecal sludge used in farming to dry before being used as a human manure. The World Health Organization has developed guidelines for safe water use.

## WATER SCARCITY

Fifty years ago, the common perception was that water was an infinite resource. At that time, there were fewer than half the current number of people on the planet. People were not as wealthy as today, consumed fewer calories and ate less meat, so less water was needed to produce their food. They required a third of the volume of water we presently take from rivers. Today, the competition for water resources is much more intense. This is because there are now more than seven billion people on the planet, their consumption of water-thirsty meat and vegetables is rising, and there is increasing competition for water from industry, urbanisation and biofuel crops. To avoid a global water crisis, farmers will have to strive to increase productivity to meet growing demands for food, while industry and cities find ways to use water more efficiently.

Successful agriculture is dependent upon farmers having sufficient access to water. However, water scarcity is already a critical constraint to farming in many parts of the world. With regards to agriculture, the World Bank targets food production and water management as an increasingly global issue that is fostering a growing debate. Physical water scarcity is where there is not enough water to meet all demands, including that needed for ecosystems to function effectively. Arid regions frequently suffer from physical water scarcity. It also occurs where water seems abundant but where resources are over-committed.

This can happen where there is overdevelopment of hydraulic infrastructure, usually for irrigation. Symptoms of physical water scarcity include environmental degradation and declining groundwater. Economic scarcity, meanwhile, is caused by a lack of investment in water or insufficient human capacity to satisfy the demand for water. Symptoms of economic water scarcity include a lack of infrastructure, with people often having to fetch water from rivers for domestic and agricultural uses. Some 2.8 billion people currently live in water-scarce areas.

## How an in-ground Irrigation System Works

Most commercial and residential irrigation systems are "in ground" systems, which means that everything is buried in the ground. With the pipes, sprinklers, emitters (drippers), and irrigation valves being hidden, it makes for a cleaner, more presentable landscape without garden hoses or other items having to be moved around manually. This does, however, create some drawbacks in the maintenance of a completely buried system.

## Controllers, Zones, and Valves

Most irrigation systems are divided into zones. A zone is a single irrigation valve and one or a group of drippers or sprinklers that are connected by pipes or tubes. Irrigation systems are divided into zones because there is usually not enough pressure and available flow to run sprinklers for an entire yard or sports field at once.

Each zone has a solenoid valve on it that is controlled via wire by an irrigation controller. The irrigation controller is either a mechanical (now the "dinosaur" type) or electrical device that signals a zone to turn on at a specific time and keeps it on for a specified amount of time. "Smart Controller" is a recent term used to describe a controller that is capable of adjusting the watering time by itself in response to current environmental conditions. The smart controller determines current conditions by means of historic weather data for the local area, a soil moisture sensors (water potential or water content), rain sensor, or in more sophisticated systems satellite feed weather station, or a combination of these.

## Emitters and Sprinklers

When a zone comes on, the water flows through the lateral lines and ultimately ends up at the irrigation emitter (drip) or sprinkler heads. Many sprinklers have pipe thread inlets on the bottom of them which allows a fitting and the pipe to be attached to them. The sprinklers are usually installed with the top of the head flush with the ground surface. When the water is pressurized, the head will pop up out of the ground and water the desired area until the valve closes and shuts off that zone. Once there is no more water pressure in the lateral line, the sprinkler head will retract back into the ground. Emitters are generally laid on the soil surface or buried a few inches to reduce evaporation losses.

### Problems in Irrigation

*Irrigation can lead to a number of problems*:

- Competition for surface water rights.
- Depletion of underground aquifers.
- Ground subsidence (*e.g.* New Orleans, Louisiana)
- Underirrigation or irrigation giving only just enough water for the plant (*e.g.* in drip line irrigation) gives poor soil salinity control which leads to increased soil salinity with consequent build up of toxic salts on soil surface in areas with high evaporation. This requires either leaching to remove these salts and a method of drainage to carry the salts away. When using drip lines, the leaching is best done regularly at certain intervals (with only a slight excess of water), so that the salt is flushed back under the plant's roots.
- Overirrigation because of poor distribution uniformity or management wastes water, chemicals, and may lead to water pollution.
- Deep drainage (from over-irrigation) may result in rising water tables which in some instances will lead to problems of irrigation salinity requiring watertable control by some form of subsurface land drainage.
- Irrigation with saline or high-sodium water may damage soil structure owing to the formation of alkaline soil

## CONSISTS OF IRRIGATION SYSTEM

The irrigation system consists of a (main) intake structure or (main) pumping station, a conveyance system, a distribution system, a field application system and a drainage system.

The (main) intake structure, or (main) pumping station, directs water from the source of supply, such as a reservoir or a river, into the irrigation system. The conveyance system assures the transport of water from the main intake structure or main pumping station up to the field ditches. The distribution system assures the transport of water through field ditches to the irrigated fields. The field application system assures the transport of water within the fields. The drainage system removes the excess water (caused by rainfall and/ or irrigation) from the fields.

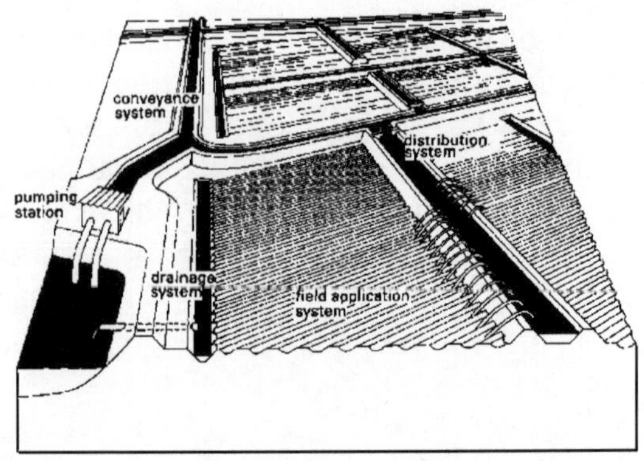


Fig. An Irrigation System

## MAIN INTAKE STRUCTURE AND PUMPING STATION

### Main Intake Structure

The intake structure is built at the entry to the irrigation system. Its purpose is to direct water from the original source of supply (lake, river, reservoir etc.) into the irrigation system.

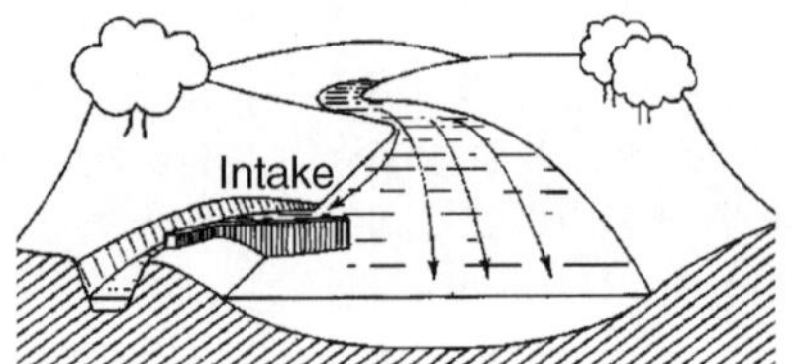


Fig. An Intake Structure

### Pumping Station

In some cases, the irrigation water source lies below the level of the irrigated fields. Then a pump must be used to supply water to the irrigation system.

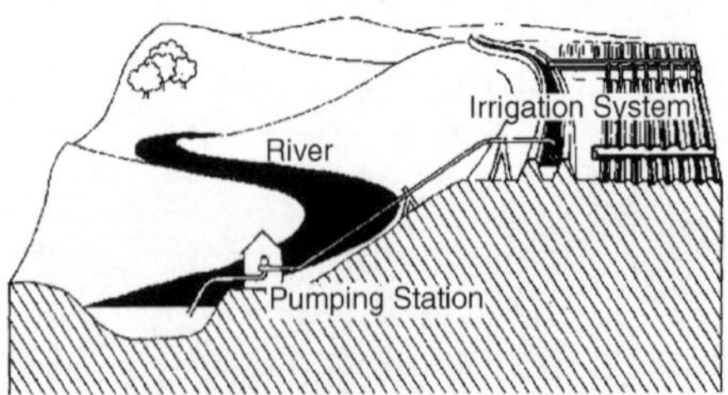


Fig. A Pumping Station

There are several types of pumps, but the most commonly used in irrigation is the centrifugal pump.

The centrifugal pump consists of a case in which an element, called an impeller, rotates driven by a motor. Water enters the case at the centre, through the suction pipe. The water is immediately caught by the rapidly rotating impeller and expelled through the discharge pipe. The centrifugal pump will only operate when the case is completely filled with water.

## CONVEYANCE AND DISTRIBUTION SYSTEM

The conveyance and distribution systems consist of canals transporting the water through the whole irrigation system. Canal structures are required for the control and measurement of the water flow.

## OPEN CANALS

An open canal, channel, or ditch, is an open waterway whose purpose is to carry water from one place to another. Channels and canals refer to main waterways supplying water to one or more farms. Field ditches have smaller dimensions and convey water from the farm entrance to the irrigated fields.

## CANAL CHARACTERISTICS

According to the shape of their cross-section, canals are called rectangular (a), triangular (b), trapezoidal (c), circular (d), parabolic (e) and irregular or natural (f).

The most commonly used canal cross-section in irrigation and drainage, is the trapezoidal cross-section. For the purposes of this publication, only this type of canal will be considered. The typical cross-section of a trapezoidal canal.

The freeboard of the canal is the height of the bank above the highest water level anticipated. It is required to guard against overtopping by waves or unexpected rises in the water level. The side slope of the canal is expressed as ratio, namely the vertical distance or height to the horizontal distance or width. For example, if the side slope of the canal has a ratio of 1:2 (one to two), this means that the horizontal distance (w) is two times the vertical distance (h).

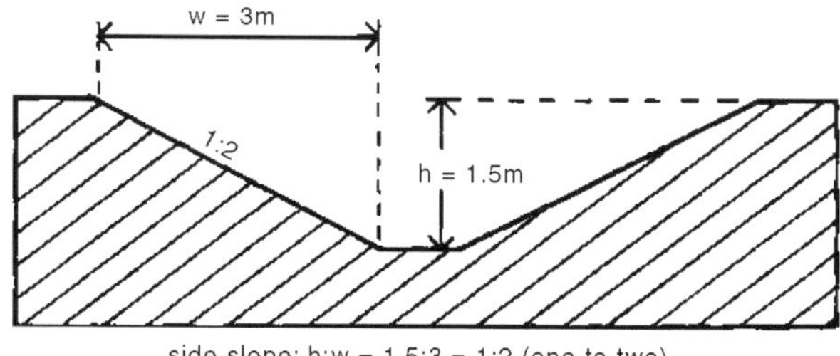


**Fig.** A Side Slope of 1:2 (One to Two)

The bottom slope of the canal does not appear on the drawing of the cross-section but on the longitudinal section. It is commonly expressed in per cent or per mil.

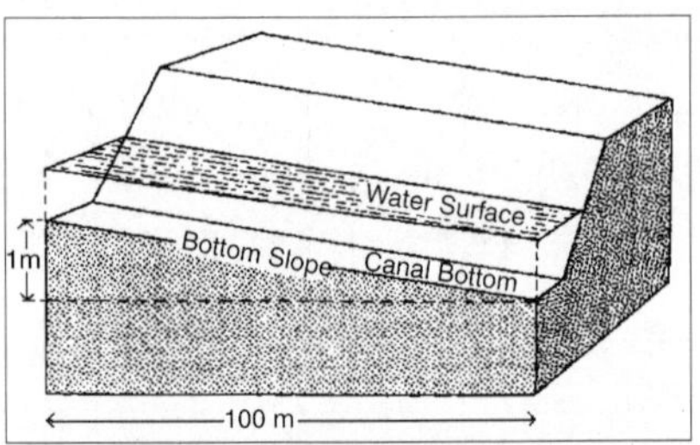


**Fig.** A Bottom Slope of a Canal

*An example of the calculation of the bottom slope of a canal is given below:*

The bottom slope (%) =

$$\frac{\text{Height difference (metres)}}{\text{Horizontal distance (metres)}} \times 100$$

$$\frac{1\text{m}}{100\text{m}} \times 100 = 1\%$$

or,

$$\text{The bottom slope (\%)} = \frac{\text{Height difference (metres)}}{\text{Horizontal distance (metres)}} \times 1000$$

$$= \frac{1\text{m}}{100\text{m}} \times 1000 = 10‰$$

## EARTHEN CANALS

Earthen canals are simply dug in the ground and the bank is made up from the removed earth. The disadvantages of earthen canals are the risk of the side slopes collapsing and the water loss due to seepage. They also require continuous maintenance in order to control weed growth and to repair damage done by livestock and rodent.

## LINED CANALS

Earthen canals can be lined with impermeable materials to prevent excessive seepage and growth of weeds. Lining canals is also an effective way to control canal bottom and bank erosion. The materials mostly used for canal lining are concrete (in precast slabs or cast in place), brick or rock masonry and asphaltic concrete (a mixture of sand, gravel and asphalt). The construction cost is much higher than for earthen canals. Maintenance is reduced for lined canals, but skilled labour is required.

## CANAL STRUCTURES

The flow of irrigation water in the canals must always be under control. For this purpose, canal structures are required. They help regulate the flow and deliver the correct amount of water to the different branches of the system and onward to the irrigated fields. There are four main types of structures: erosion control structures, distribution control structures, crossing structures and water measurement structures.

## Erosion Control Structures

### *Canal Erosion*

Canal bottom slope and water velocity are closely related, as the following example will show.

A cardboard sheet is lifted on one side 2 cm from the ground. A small ball is placed at the edge of the lifted side of the sheet. It starts rolling downward, following the slope direction.

The sheet edge is now lifted 5 cm from the ground creating a steeper slope. The same ball placed on the top edge of the sheet rolls downward, but this time much faster. The steeper the slope, the higher the velocity of the ball.

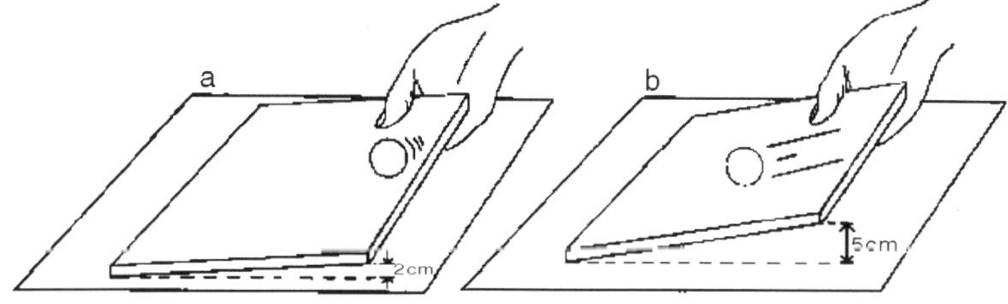


**Fig.** The Relationship between Slope and Velocity

Water poured on the top edge of the sheet reacts exactly the same as the ball. It flows downward and the steeper the slope, the higher the velocity of the flow.

Water flowing in steep canals can reach very high velocities. Soil particles along the bottom and banks of an earthen canal are then lifted, carried away by the water flow and deposited downstream where they may block the canal and silt up structures. The canal is said to be under erosion; the banks might eventually collapse.

### *Drop Structures and Chutes*

Drop structures or chutes are required to reduce the bottom slope of canals lying on steeply sloping land in order to avoid high velocity of the flow and risk of erosion. These structures permit the canal to be constructed as a series of relatively flat sections, each at a different elevation.

Drop structures take the water abruptly from a higher section of the canal to a lower one. In a chute, the water does not drop freely but is carried through a steep, lined canal section. Chutes are used where there are big differences in the elevation of the canal.

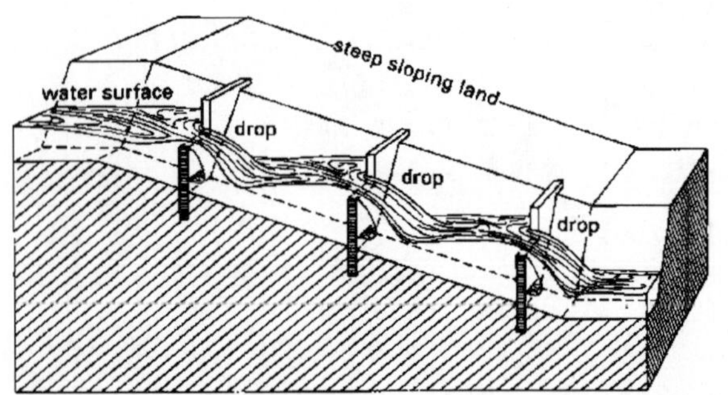


**Fig.** Longitudinal Section of a Series of Drop Structures

### *Distribution Control Structures*

Distribution control structures are required for easy and accurate water distribution within the irrigation system and on the farm. Division boxes are used to divide or direct the flow of water between two or more canals or ditches. Water enters the box through an opening on one side and flows out through openings on the other sides. These openings are equipped with gates.

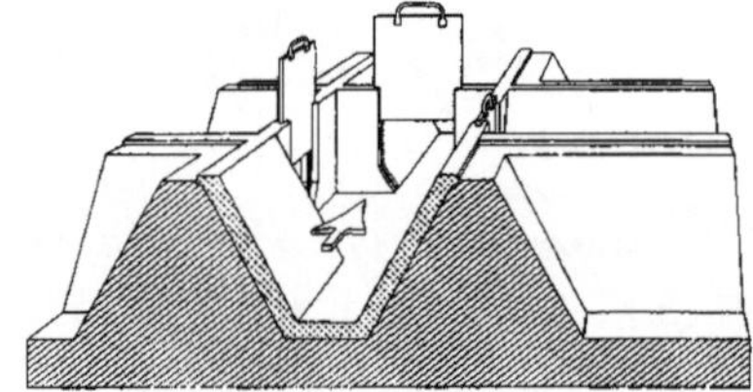

**Fig.** A Division Box with Three Gates

### *Turnouts*

Turnouts are constructed in the bank of a canal. They divert part of the water from the canal to a smaller one.

Turnouts can be concrete structures, or pipe structures.

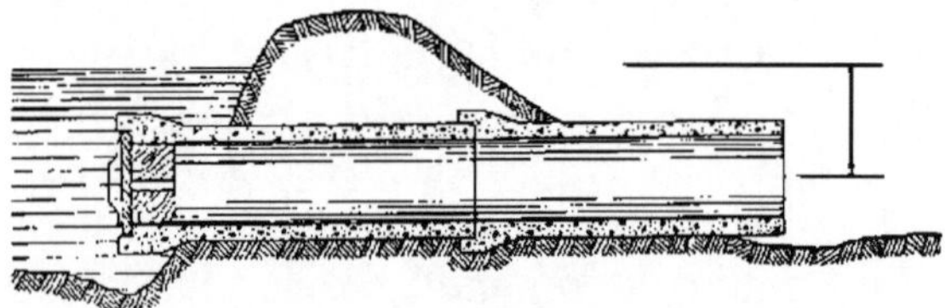

**Fig.** A Pipe Turnout

### *Checks*

To divert water from the field ditch to the field, it is often necessary to raise the water level in the ditch. Checks are structures placed across the ditch to block it temporarily and to raise the upstream water level. Checks can be permanent structures or portable.

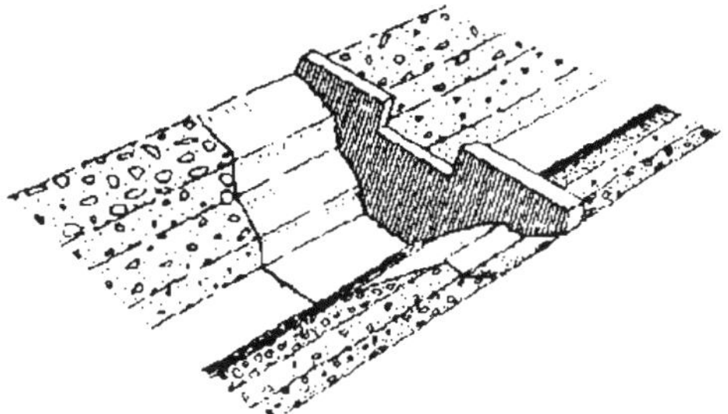

**Fig.** A Permanent Concrete Check

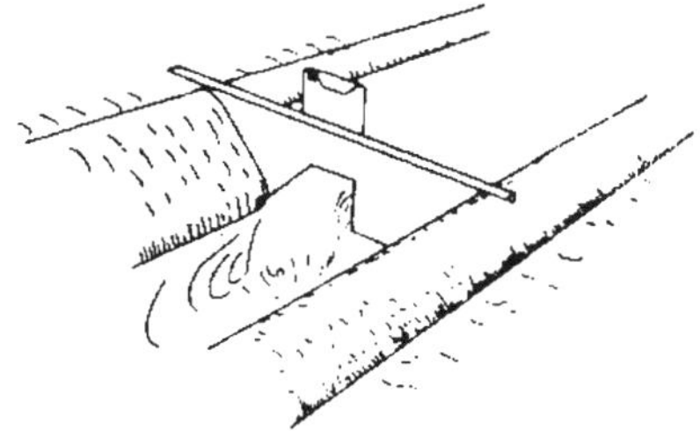

**Fig.** A Portable Metal Check

## CROSSING STRUCTURES

It is often necessary to carry irrigation water across roads, hillsides and natural depressions. Crossing structures, such as flumes, culverts and inverted siphons, are then required.

### Flumes

Flumes are used to carry irrigation water across gullies, ravines or other natural depressions.

They are open canals made of wood (bamboo), metal or concrete which often need to be supported by pillars.

### *Culverts*

Culverts are used to carry the water across roads. The structure consists of masonry or concrete headwalls at the inlet and outlet connected by a buried pipeline.

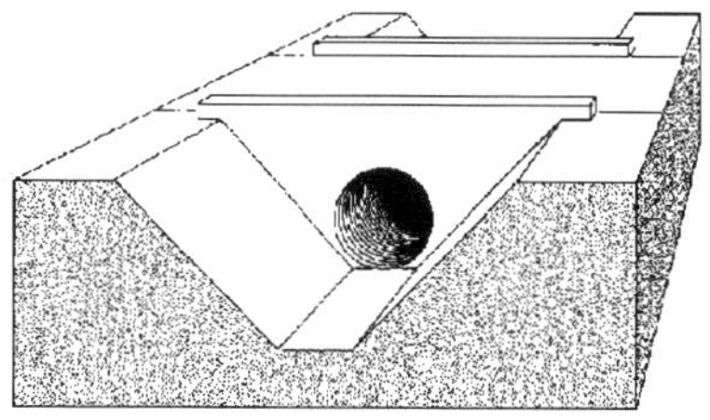

**Fig.** A Culvert

### *Inverted Siphons*

When water has to be carried across a road which is at the same level as or below the canal bottom, an inverted siphon is used instead of a culvert. The structure consists of an inlet and outlet connected by a pipeline. Inverted siphons are also used to carry water across wide depressions.

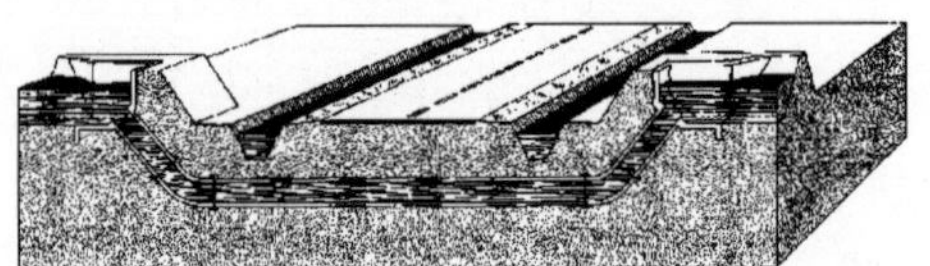

**Fig.** An Inverted Siphon

## Water Measurement Structures

The principal objective of measuring irrigation water is to permit efficient distribution and application. By measuring the flow of water, a farmer knows how much water is applied during each irrigation. In irrigation schemes where water costs are charged to the farmer, water measurement provides a basis for estimating water charges. The most commonly used water measuring structures are weirs and flumes. In these structures, the water depth is read on a scale which is part of the structure. Using this reading, the flow-rate is then computed from standard formulas or obtained from standard tables prepared specially for the structure.

### *Weirs*

In its simplest form, a weir consists of a wall of timber, metal or concrete with an opening with fixed dimensions cut in its edge. The opening, called a notch, may be rectangular, trapezoidal or triangular.

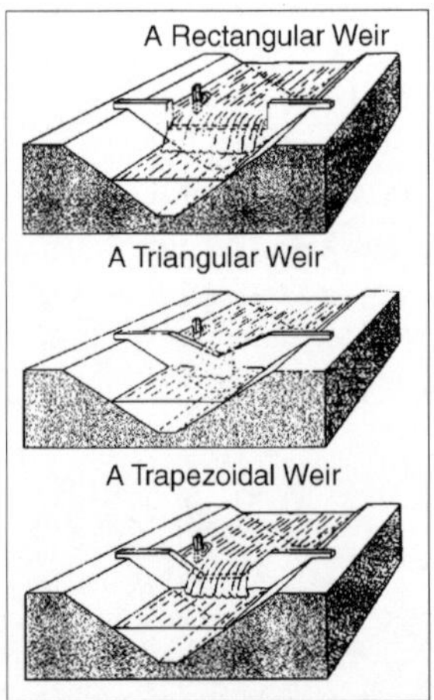


**Fig.** Some Examples of Weirs

***Parshall Flumes***

The Parshall flume consists of a metal or concrete channel structure with three main sections:

- A converging section at the upstream end, leading to;
- A constricted or throat section,
- A diverging section at the downstream end.

Depending on the flow condition (free flow or submerged flow), the water depth readings are taken on one scale only (the upstream one) or on both scales simultaneously.

**Cut-throat Flume**

The cut-throat flume is similar to the Parshall flume, but has no throat section, only converging and diverging sections. Unlike the Parshall flume, the cut-throat flume has a flat bottom. Because it is easier to construct and install, the cut-throat flume is often preferred to the Parshall flume.

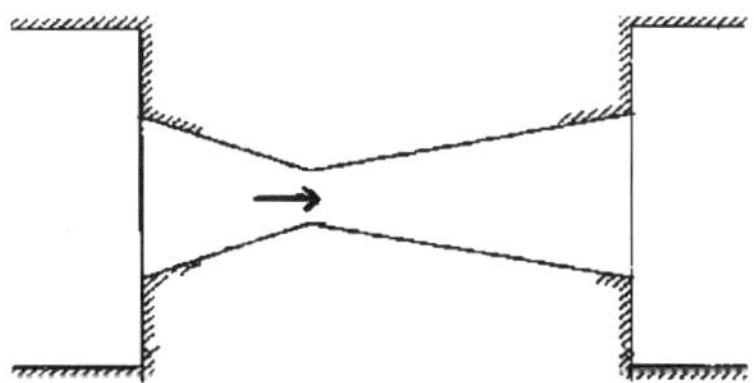

**Fig.** A Cut-throat Flume

## FIELD APPLICATION SYSTEMS

There are many methods of applying water to the field. The simplest one consists of bringing water from the source of supply, such as a well, to each plant with a bucket or a water-can.

This is a very time-consuming method and it involves quite heavy work. However, it can be used successfully to irrigate small plots of land, such as vegetable gardens, that are in the neighbourhood of a water source. More sophisticated methods of water application are used in larger irrigation systems. There are three basic methods: surface irrigation, sprinkler irrigation and drip irrigation.

**Surface Irrigation**

Surface irrigation is the application of water to the fields at ground level. Either the entire field is flooded or the water is directed into furrows or borders.

***Furrow Irrigation***

Furrows are narrow ditches dug on the field between the rows of crops. The water runs along them as it moves down the slope of the field. The water flows from the field ditch into the furrows by opening up the bank or dyke of the ditch or by means of syphons or spiles. Siphons are small curved pipes that deliver water over the ditch bank. Spiles are small pipes buried in the ditch bank.

### *Border Irrigation*

In border irrigation, the field to be irrigated is divided into strips (also called borders or borderstrips) by parallel dykes or border ridges. The water is released from the field ditch onto the border through gate structures called outlets. The water can also be released by means of siphons or spiles. The sheet of flowing water moves down the slope of the border, guided by the border ridges.

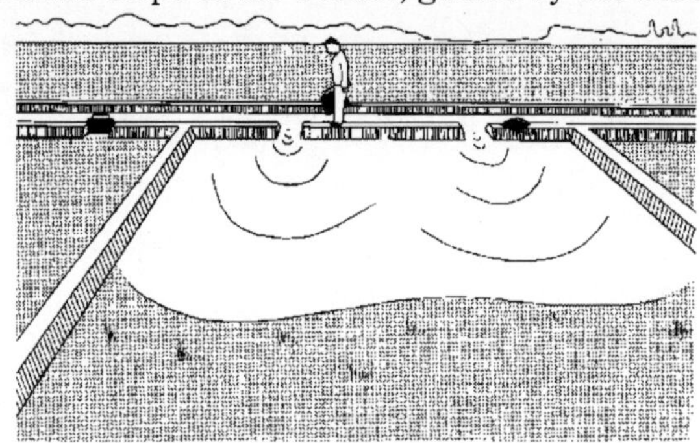

**Fig.** Border Irrigation

### *Basin Irrigation*

Basins are horizontal, flat plots of land, surrounded by small dykes or bunds. The banks prevent the water from flowing to the surrounding fields. Basin irrigation is commonly used for rice grown on flat lands or in terraces on hillsides. Trees can also be grown in basins, where one tree usually is located in the centre of a small basin.

### Sprinkler Irrigation

With sprinkler irrigation, artificial rainfall is created. The water is led to the field through a pipe system in which the water is under pressure. The spraying is accomplished by using several rotating sprinkler heads or spray nozzles or a single gun type sprinkler.

### Drip Irrigation

In drip irrigation, also called trickle irrigation, the water is led to the field through a pipe system. On the field, next to the row of plants or trees, a tube is installed. At regular intervals, near the plants or trees, a hole is made in the tube and equipped with an emitter. The water is supplied slowly, drop by drop, to the plants through these emitters.

**Drainage System**

A drainage system is necessary to remove excess water from the irrigated land. This excess water may be *e.g.* waste water from irrigation or surface run-off from rainfall. It may also include leakage or seepage water from the distribution system.

## NEED FOR DRAINAGE

During rain or irrigation, the fields become wet. The water infiltrates into the soil and is stored in its pores. When all the pores are filled with water, the soil is said to be saturated and no more water can be absorbed; when rain or irrigation continues, pools may form on the soil surface. Part of the water present in the saturated upper soil layers flows downward into deeper layers and is replaced by water infiltrating from the surface pools.

When there is no more water left on the soil surface, the downward flow continues for a while and air re-enters in the pores of the soil. This soil is not saturated anymore. However, saturation may have lasted too long for the plants' health. Plant roots require air as well as water and most plants cannot withstand saturated soil for long periods (rice is an exception). Besides damage to the crop, a very wet soil makes the use of machinery difficult, if not impossible.

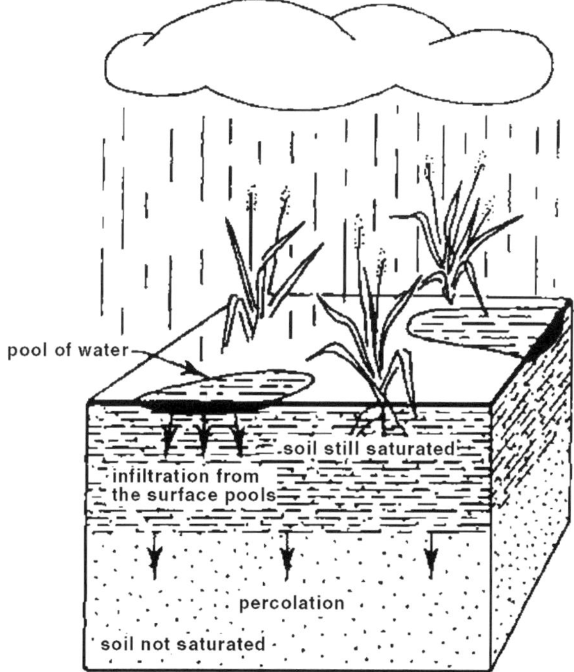


**Fig.** Water Percolates to Deeper Layers and Infiltrates from the Pools

The water flowing from the saturated soil downward to deeper layers, feeds the groundwater reservoir. As a result, the groundwater level (often called groundwater table or simply water table) rises. Following heavy rainfall or continuous over-irrigation, the groundwater table may even reach and saturate part of the rootzone.

Again, if this situation lasts too long, the plants may suffer. Measures to control the rise of the water table are thus necessary. The removal of excess water either from the ground surface or from the rootzone, is called drainage. Excess water may be caused by rainfall or by using too much irrigation water, but may also have other origins such as canal seepage or floods.

In very dry areas there is often accumulation of salts in the soil. Most crops do not grow well on salty soil. Salts can be washed out by percolating irrigation water through the rootzone of the crops.

To achieve sufficient percolation, farmers will apply more water to the field than the crops need. But the salty percolation water will cause the water table to rise. Drainage to control the water table, therefore, also serves to control the salinity of the soil.

## DIFFERENT TYPES OF DRAINAGE

Drainage can be either natural or artificial. Many areas have some natural drainage; this means that excess water flows from the farmers' fields to swamps or to lakes and rivers. Natural drainage, however, is often inadequate and artificial or man-made drainage is required.

There are two types of artificial drainage: Surface drainage and subsurface drainage.

### Surface Drainage

Surface drainage is the removal of excess water from the surface of the land. This is normally accomplished by shallow ditches, also called open drains. The shallow ditches discharge into larger and deeper collector drains. In order to facilitate the flow of excess water towards the drains, the field is given an artificial slope by means of land grading.

### Subsurface Drainage

Subsurface drainage is the removal of water from the rootzone. It is accomplished by deep open drains or buried pipe drains.

#### *Pipe Drains*

Pipe drains are buried pipes with openings through which the soil water can enter. The pipes convey the water to a collector drain.

Drain pipes are made of clay, concrete or plastic. They are usually placed in trenches by machines. In clay and concrete pipes (usually 30 cm long and 5 - 10 cm in diameter) drainage water enters the pipes through the joints. Flexible plastic drains are much longer (up to 200 m) and the water enters through perforations distributed over the entire length of the pipe. Open drains use land that otherwise could be used for crops. They restrict the use of machines. They also require a large number of bridges and culverts for road crossings and access to the fields. Open drains require frequent maintenance (weed control, repairs, etc.).

In contrast to open drains, buried pipes cause no loss of cultivable land and maintenance requirements are very limited. The installation costs, however, of pipe drains may be higher due to the materials, the equipment and the skilled manpower involved.

***Deep Open Drains***

The excess water from the rootzone flows into the open drains. The disadvantage of this type of subsurface drainage is that it makes the use of machinery difficult.

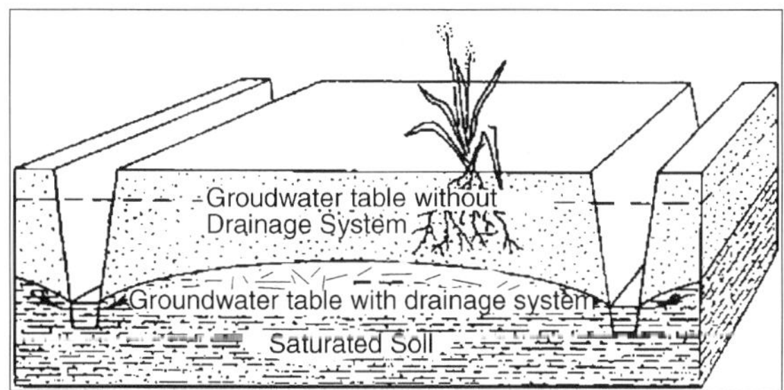


**Fig.** Control of the Groundwater Table by Means of Deep Open Drains

## SOILS SALINIZATION

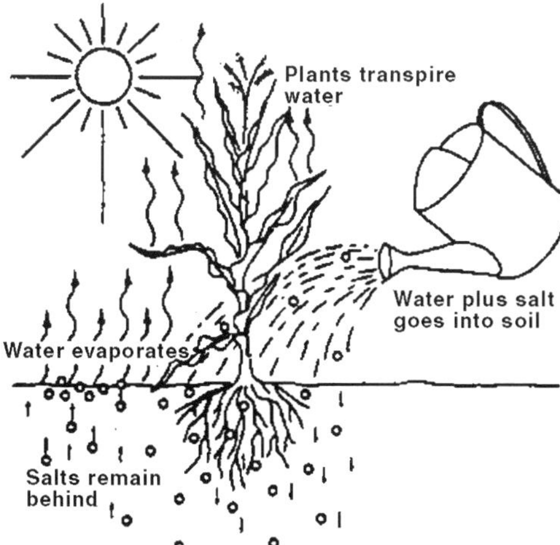


**Fig.** Salinization, Caused by Salty Irrigation Water

A soil may be rich in salts because the parent rock from which it was formed contains salts. Sea water is another source of salts in low-lying areas along the coast. A very common source of salts in irrigated soils is the irrigation water itself. Most irrigation waters contain some salts. After irrigation, the water added to the soil is used by the crop or evaporates directly from the moist soil. The salt, however, is left behind in the soil. If not removed, it accumulates in the soil; this process is called salinization. Very salty soils are sometimes recognizable by a white layer of dry salt on the soil surface.

Salty groundwater may also contribute to salinization. When the water table rises (*e.g.* following irrigation in the absence of proper drainage), the salty groundwater may reach the upper soil layers and, thus, supply salts to the rootzone.

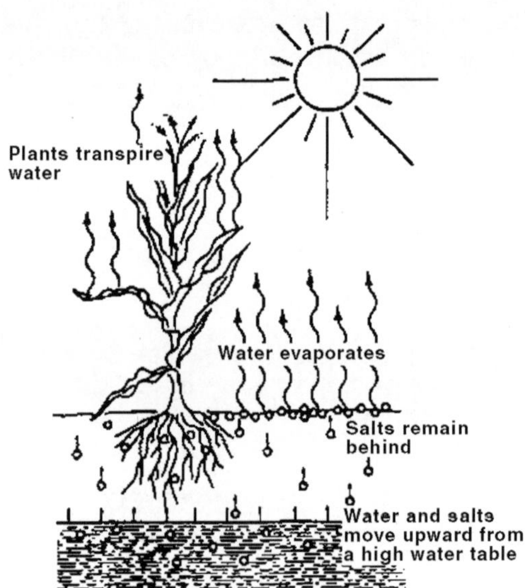


**Fig.** Salinization, Caused by a High

Soils that contain a harmful amount of salt are often referred to as salty or saline soils. Soil, or water, that has a high content of salt is said to have a high salinity.

## WATER SALINITY

Water salinity is the amount of salt contained in the water. It is also called the "salt concentration" and may be expressed in grams of salt per litre of water (grams/litre or g/l) or in milligrams per litre (which is the same as parts per million, p.p.m).

However, the salinity of both water and soil is easily measured by means of an electrical device. It is then expressed in terms of electrical conductivity: millimhos/cm or micromhos/cm. A salt concentration of 1 gram per litre is about 1.5 millimhos/cm. Thus a concentration of 3 grams per litre will be about the same as 4.5 millimhos/cm.

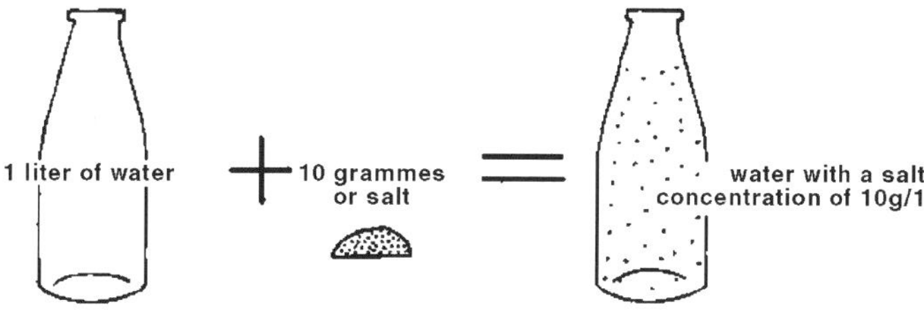


**Fig.** A Salt Concentration of 10 g/l

## Soil Salinity

The salt concentration in the water extracted from a saturated soil (called saturation extract) defines the salinity of this soil. If this water contains less than 3 grams of salt per litre, the soil is said to be non- saline. If the salt concentration of the saturation extract contains more than 12 g/l, the soil is said to be highly saline.

**Table. Salt Concentration of the Soil Water (Saturation Extract)**

| *In g/l* | *In Millimhos/cm* | *Salinity* |
|---|---|---|
| 0 - 3 | 0 - 4.5 | non- saline |
| 3 - 6 | 4.5 - 9 | slightly saline |
| 6 - 12 | 9 - 18 | medium saline |
| more than 12 | more than 18 | highly saline |

## CROPS AND SALINE SOILS

Most crops do not grow well on soils that contain salts. One reason is that salt causes a reduction in the rate and amount of water that the plant roots can take up from the soil. Also, some salts are toxic to plants when present in high concentration. Some plants are more tolerant to a high salt concentration than others. Some examples are given in the following table:

| *Highly Tolerant* | *Moderately Tolerant* | *Sensitive* |
|---|---|---|
| Date palm | Wheat | Red clover |
| Barley | Tomato | Peas |
| Sugarbeet | Oats | Beans |
| Cotton | Alfalfa | Sugarcane |
| Asparagus | Rice | Pear |
| Spinach | Maize | Apple |
| | Flax | Orange |
| | Potatoes | Prune |
| | Carrot | Plum |
| | Onion | Almond |
| | Cucumber | Apricot |
| | Pomegranate | Peach |
| | Fig | |

| | |
|---|---|
| | Olive |
| | Grape |

The highly tolerant crops can withstand a salt concentration of the saturation extract up to 10 g/l. The moderately tolerant crops can withstand salt concentration up to 5 g/l. The limit of the sensitive group is about 2.5 g/l.

### Sodicity

Salty soils usually contain several types of salt. One of these is sodium salt. Where the concentration of sodium salts is high relative to other types of salt, a sodic soil may develop. Sodic soils are characterized by a poor soil structure: they have a low infiltration rate, they are poorly aerated and difficult to cultivate. Thus, sodic soils adversely affect the plants' growth.

## IMPROVEMENT OF SALINE AND SODIC SOILS

Numerous areas in the world are naturally saline or sodic or have become saline due to improper irrigation practices. Crop growth on many of these is poor. However, their productivity can be improved by a number of measures.

### Improvement of Saline Soils

Improvement of a saline soil implies the reduction of the salt concentration of the soil to a level that is not harmful to the crops. To that end, more water is applied to the field than is required for crop growth. This additional water infiltrates into the soil and percolates through the rootzone. During percolation, it takes up part of the salts in the soil and takes these along to deeper soil layers. In fact, the water washes the salts out of the rootzone. This washing process is called leaching.

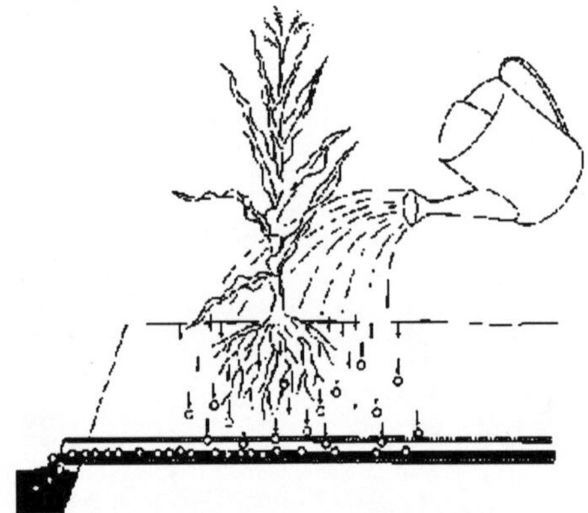

**Fig.** Leaching of Salts

The additional water required for leaching must be removed from the rootzone by means of a subsurface drainage system. If not removed, it could cause a rise of the groundwater table which would bring the salts back into the rootzone. Thus, improvement of saline soils includes, essentially, leaching and sub-surface drainage.

## PREVENTION OF SALINIZATION

Soils will become salty if salts are allowed to accumulate. Proper irrigation management and adequate drainage are not only important measures for the improvement of salty soils, they are also essential for the prevention of salinization.

### Irrigation Water Quality

The suitability of water for irrigation depends on the amount and the type of salt the irrigation water contains. The higher the salt concentration of the irrigation water, the greater the risk of salinization. The following Table gives an idea of the risk of salinization:

| *Salt Concentration of the Irrigation water in g/l* | *Soil Salinization risk* | *Restriction on use* |
|---|---|---|
| Less than 0.5 g/l | No risk | No restriction on its use |
| 0.5 - 2 g/l | Slight to moderate risk | should be used with approp riate water management practices |
| More than 2 g/l | High Risk | Not generally advised for use unless consulted with specialists |

The type of salt in the irrigation water will influence the risk of developing sodicity: the higher the concentration of sodium present in the irrigation water (particularly compared to other soils), the higher the risk.

### Irrigation Management and Drainage

Irrigation systems are never fully efficient. Some water is always lost in canals and on the farmers' fields. Part of this seeps into the soil. While this will help leach salt out of the rootzone, it will also contribute to a rise of the water table; a high water table is risky because it may cause the salts to return to the rootzone. Therefore, both the water losses and the water table must be strictly controlled. This requires careful management of the irrigation system and a good subsurface drainage system.

## AGRICULTURAL WATER CRISIS

The agriculture development in China relies more and more heavily on water sources. Based on the analysis of the existing problems and the improvement of society, science and technology, rational development and high efficient use of agricultural water resources is unique strategy accounting for water crisis involving in grain supply to 1.6 billion Chinese in the mid-21st century. Water crisis is affected by many factors. The following countermeasures should be taken:

## ENERGETICALLY INCREASING INPUT TO THE CONSTRUCTION

While the development of water saving in agriculture, the input to the construction of water conservancy infrastructure should be increased so as to develop various water source for irrigation.

Some important water projects for the development of national economy should be constructed to change the status of water shortage in the north part of China and the coastal regions, to improve the production condition of the important production bases of grain and cotton, and to promote the agriculture development.

Developing rain irrigation is very important not only to the agriculture development on rain-fed land, but also to the high efficient use of irrigation water and the relaxation of contradiction between the demand and the supply. The utilization degree of rain and flood during flood season should be raised by structure and non-structure measures in order to combine flood control with generating benefit, *i.e.* utilizing resources while controlling disaster. It will play an outstanding role in relaxation of water crisis and flood disaster. The treatment and reuse of wastewater and low quality water will not only relax the water crisis in irrigation, but also improve and protect the ecological environment.

### Vigorously Developing High Efficient Use of Agricultural Water Use

At present, the production efficiency of irrigation water is still less than 40 per cent, accordingly which decrease the utilization efficiency of rain and irrigation water and worsen the water crisis. The water production efficiency can be improved by taking synthesis agricultural measures to decrease crop's invalid water consumption. These agricultural measures include: planting the crops and breed with the characteristics of low water consumption and high production; improving utilization ratio of soil water and protecting soil moisture; high efficient allocation and control of water and fertilizer; covering planting; improving crop's drought resistance and water utilization efficiency by chemical measures.

The development of high efficient use of agricultural water use relies not only on technology, but also on the following sufficient studies: studying and establishing the strategy and target of the development of agricultural water saving; study on the basic gist putting forward the general target and task of agricultural water saving; for different stage and region, study on the requirement on agricultural water saving by national economy development, the national economic base supporting the development of agricultural water saving, farmer's income, investment ability and interesting, as well as the development strategy and technologic principle; for different stage, study on the mechanism prompting agricultural water saving and its change and control measures; study on the stage of industrialization of agricultural water saving.

The development of high efficient use of agricultural water use must be sustainable and accord with the following three criterions: firstly, sustainablity–can effectively protect, rationally and high efficiently utilize water resources and make the utilization sustainable; Secondly, efficiency-can markedly raise economic efficiency, including water saving, production increase, land saving, energy saving, ecological environmental benefit, profit of non-agricultural capital input, and so on; Finally, high scientific and technological characteristic-should be scientific and technologic and can fully meet the requirement on high efficient use of agricultural water use at different development stage.

High Efficient use of agricultural water use is the foundation supporting the sustainable development of agricultural and the sustainable utilization of water resources.

**Energetically Improving and Protecting**

The simple improvement with protection target should be transferred to the improvement with exploitation target in order to combine tightly ecological benefit with economic benefit. The exploitation should be combined with and promotes the improvement. Strengthening the construction of water conservancy infrastructure on farmland in order to raise the farmland's ability of resisting water-logging and saline disasters, and rationally utilize the water resources. The pollution of the agricultural water resources should be effectively prevented and improved by the rational and effective use of fertilizer and bad water source.

**Intensifying the Management For High Efficient**

It is recognized in the World that 50 per cent of potential of irrigation water saving is from the improvement of management. There is a very big gap in management plane of water resources between China and the developed countries.

Firstly, the policies and codes prompting high efficient use of water resources are not perfect. There is short of mechanism and management measures encouraging high efficient use of agricultural water use. Secondly, the technical system supporting high efficient use of water resources has not been established. The potential of water resources far has not been fully developed. Thirdly, there is short of engineering measure which can support the improvement of management, because of the low project standards of many agricultural water projects.

*Therefore, to resolve the water crisis, the water management must be changed from an extensive pattern to an intensive pattern, including:*

- Establishing the management system supporting rational and high efficient use of water resources to make the unified management of water resources available.

- Establishing the price system prompting rational use of water resources to form a sound market mechanism for the management of water use.
- Establishing the guarantee system prompting high efficient use of agricultural water use, including investment mechanism.

**Strengthening Scientific Study and Research**

With the contradiction between the demand and the supply is day and day outstanding and the national economy is developing, many technical problems related to the high efficient water use are waiting for resolving and new technical problems appear in succession. All the problems need to be profoundly studied. The technical integration and innovate system supporting the high efficient use of agricultural water use should be established. The science and education should benefit the development of water resources. The development of series products of high efficient water use should be strengthened. The efficiency of water use should be raised.

**Reforming the Mechanism of Water Management**

The technical reform and management system reform of irrigation districts must be pressed on to reach the target of high efficient use of water resources and solidify the agricultural production on irrigated area on which the crop's production account for third-fourth of the entire country. The technical reform of irrigation districts is basis of high efficient water use, and the management system reform of irrigation districts is its assurance. The management system reform of irrigation districts will benefit not only for the exertion and increase of value of state's capital, but also for the sustainable development of irrigation agriculture and high efficient water use, even for food security. So it is one of the important measures accounting for the water crisis, and is also a key component of the prosecution and management system reform in agriculture and rural. This reform should be in line with the development of market agriculture and integration of prosecution to be one part of the socialism market economy.

Because of the limit of natural condition, immense requirement of the society and economy development, weak economic base, and so on, the sustainable development of agriculture is facing severe water crisis. With the population increase, people's living standard raise and cultivated land decrease, the press on cultivated land, specially on irrigated cultivated land, will be more and more heavy. Water will dominate the fortune of China's agriculture and the survival and development of 1.6 billion Chinese in the 21st century. Based on the analysis of water crisis, social practice and the advancement of science and technology, the essential measure accounting for the water crisis faced by the sustainable development of agriculture is high efficient use of agricultural water use. The raise of water use efficiency and water production efficiency is key strategic measure related to the survival and development of 1.6 billion Chinese in the 21st century.

The development of agriculture and water is a huge system engineering. The water crisis must be roundly resolved from system viewpoint. It should be realised the water's complexity. Don't keep eyes only on water. The problem of water resources should be roundly resolved from system viewpoint and by means of carrying out the comprehensive prevention and cure strategy, to reach effect which is more than the addition of single strategy's effect.

The utilization efficiency and production efficiency of water resources must be raised by means of system engineering and carrying out the comprehensive measures including water projects, agricultural biologic measure, modern management technology, information engineering technology and meteorology, to basically resolve the contradiction between water supply and demand, and to support the rapid development of society and economy in China in the 21st century.

## WATER CRISIS LIMITS THE AGRICULTURE

Since the founding of the People's Republic of China, a large amount of water conservancy infrastructure have been constructed and strongly encouraged the development of the industry and agriculture and urban construction, resisted flood, protected environment, and obviously raised the people's living standards. With the increase in population and rapid development of the society and economy, the water supply obviously cannot meet the water demand and resulted in water shortage.

According to statistics, drought disaster bring with more serious affection than flood disaster. Therefore, a critical problem facing by the development of China's society and economy is how to satisfy the water demand of society and economy.

*Water shortage resulted in:*

- The production from rain-fed land which area is more than 50 per cent of total cultivated area in China has to rely on climate, and is low and unstable. Consequently, the agricultural development has badly been limited.
- The water shortage resulted from water pollution is also remarkable.
- In the northern part of China, a series of the ecological and environmental problems occurred and day and day worsened, including drying up of rivers, bad overdraft of groundwater, decline of groundwater table in broad area, and so on.
- Large amounts of wastewater and sewage without treatment have

been directly and indirectly used for irrigation, specially in the northern part of China, and gradually become an important component part of agricultural water use. The use of wastewater and sewage without treatment resulted in the pollution of soil water in farmland and groundwater and the excess content of pollutants in agricultural products which are harmful to the people's health. This kind of harm is very severe.

Except for the water shortage and the water environmental problem, China's water crisis still includes the frequent flood disaster and the low capability of resisting natural disasters. Flood disaster is still a serious danger to the sustainable development of China's agriculture.

## WATER CRISIS IN THE 21ST CENTURY IN CHINA

*The water demand and supply in the 21st century in China will face:*

- Due to too fast increase in water demand and limited increase in water supply, the lag between demand and supply will further be widened;
- Because of the existing water source drying up, water projects aging and inadequate maintenance, and so on, the water supply sharp decline;
- In the northern part of China, the available water supply obviously decrease with deep development of water resources and the water consumption increase in river basin;
- The warmer and warmer climate may aggravate water crisis;
- With industrial and domestic water use increase, water pollution will more and more seriously threaten water supply and safety of water use;
- People don't fully understand the seriousness of water crisis, thereby water projects aren't enough and management is backward. Water supply lag water demand with extravagant water use. The low water use efficiency and water production efficiency further worsen the water crisis.

The 21st century will be an era of the great development of economy, and an era which China's population will reach the peak of 1.6 billion. Water is an important constraining factor for the economy development. If the problem of water shortage cannot be resolved, the sustainable development of society and economy must be seriously imperiled.

*Moreover, the water crisis mainly affect the agriculture development by:*

- Competition for water among regions, between industry and agriculture, between urban and rural, will continue for a long time and be more serious.
- Competition for water between agriculture and environment will stand out, specially in the northern part of China.

- With the development of economy and the raise of people's living standards, the water demand by other agriculture, including forestry, animal husbandry, fishery and subsidiary, must extremely increase. Accordingly, competition for water within agriculture sector will be more outstanding.

## THE DEVELOPMENT OF IRRIGATION AREA AND AGRICULTURAL WATER USE

Rural economy rapidly developed and grain production steady increased, but the increment of agricultural water use was limited. Though the actual irrigation water use is affected by many factors including yearly hydrologic and meteorologic conditions, the general trend of water use in the whole country still can be concluded.

The decrease of irrigation water use was mainly due to effective agricultural water saving and shortage of water supply for agriculture in water scarcity regions.

With the development and popularization of agricultural water saving, the efficiency of agricultural water use is being raised. The actual irrigation area increased by about 8,200 thousand $hm^2$ from 1980 to 1999, but the change of irrigation water use was tiny. In general, the average water use per $hm^2$ decreased from 8,750 $m^3$ in 1980 to 7,270 $m^3$ in 1999. The water use of per unit area decreased by 17 per cent. These agricultural statistics are shown in Table.

The water use by forest, herd and fishery fast increased while irrigation water use decreasing. Sequentially, the water demand by ecological environment had better been met.

**Table. Irrigation Area and Water Use**

| *Year* | *Irrigation Water Use ($B\ m^3$)* | *Actual Irrigation Area ($10^3\ hm^2$)* | *Unit water use ($m^3/hm^2$)* |
|---|---|---|---|
| 1980 | 358.1 | 40 920 | 8 750 |
| 1999 | 356.4 | 49 090 | 7 260 |
| Increment | –1.7 | 8 170 | –1 490 |

## WATER DEMAND

Based on the analysis of the relationship between water resources and the sustainable development of national economy, the characteristic of water resources, factors of water resources rational allocation supporting the sustainable development, status of water use and supply, as well as the existing problems, the water demand before the mid-21st century have been predicted.

The prediction on water demand by industry, agriculture and domesticity in different target years are shown in Table.

**Table. Prediction on Water Demand (B $m^3$)**

| Year | Urban | Rural | Industry | Irrigation | Other Agriculture | Total |
|---|---|---|---|---|---|---|
| 2010 | 40.5 | 30.2 | 149.8 | 387.9 | 34.0 | 642.4 |
| 2030 | 64.1 | 30.9 | 191.1 | 387.2 | 38.5 | 711.8 |
| 2050 | 81.5 | 30.6 | 199.8 | 377.5 | 42.5 | 731.9 |

Table indicates the basic increasing trend of water use: faster increasing of industrial and domestic water use, little increasing of agricultural water use, steady growth of proportion of industrial and domestic water use in total water use, decrease of proportion of agricultural water use. However, because of natural condition, the agricultural development in China has to rely on irrigation.

So the proportion of irrigation water use in total national economy water use always is highest among the sectors. According to the status and the development trend, three main problems, such as flood disaster, water shortage and water environment worsen, especially the water shortage will more and more heavily limit the development of agriculture, society and economy.

How to resolve the contradiction between the sustainable development of agriculture and water supply is one of the focus related to the sustainable development of society and economy.

## CURVE NUMBERS, RECENT DEVELOPMENTS

The Curve Number procedure of the U.S. Dept. of Agriculture, Natural Resources Conservation Service (NRCS) (formerly Soil Conservation Service, SCS) has elicited questions and concern since its conception. This arises, for the most part, because users read into the procedure what they wish was covered. The actual intent of the procedure is often disregarded. Too, the basic reference for Curve Numbers, the National Engineering Handbook of the SCS, has been revised several times, not always by individuals or committees that understood the significance of their statements. The Natural Resources Conservation Service and the Agricultural Research Service, both agencies of the U.S. Department of Agriculture, formed a joint work group to assess the state of the Curve Number procedure and to chart its future development.

*The joint work group recognized three distinctly different modes of application for Curve Numbers:*

1. Determination of run-off volume of a given return period, given total event rainfall for that return period;
2. Determine direct run-off for individual events, explaining the variability from event to event, as used in continuous simulation models;
3. Determine infiltration rates for short time intervals as used with unit hydrograph development of flood hydrographs.

The first mode of application represents the historical basis of the procedure, so receives the most attention. Use as a surrogate for an infiltration is very common and follows from the historical basis, so must be considered. The application in continuous simulation models is an extension beyond the scope of the committee.

Discussions within the committee made it apparent that a portion of the difficulty surrounding the procedure was attributable to the presentation of the procedure in the National Engineering Handbook. The first task then was to rewrite those portions of the Handbook pertaining to the procedure. Problems identified ranged from incorrect and misleading statements to incomplete documentation. For example, it was incorrectly stated that S includes Ia, whereas it can be shown mathematically that S does not include Ia.

Fortunately, this is only significant for continuous simulation. Another example was a table that related antecedent rainfall to antecedent moisture condition (AMC). This was not intended to have nationwide application, though it was treated as such. Folklore concerning Curve Numbers could also be attributed to problems with documentation. A folklore example is that the Curve Number Run-off Equation is an infiltration equation.

*In rewriting the Curve Number portions of the Handbook, the work group agreed that:*

- Committee must Abelieve in≅ concepts expressed;
- References will be included if possible; and
- Results must be technically defensible.

*The results of the rewrite include such items as:*

- Reference to Antecedent Moisture Condition (AMC) was removed. Variability is incorporated by considering the curve number as a random variable and the AMC–I and AMC–III conditions as bounds on the distribution.
- Reiteration of desirability of locally determined curve numbers. This was part of the original documentation but tended to be neglected.
- Explicit expression of Curve Number run-off equation as a transformation of rainfall frequency distribution to run-off frequency distribution. This was demonstrated in the original documentation but again was often neglected.
- Expression of AMC-I and AMC–III as measures of dispersion about the central tendency (AMC II). This is a corollary of treating the CN as a random variable.
- Mathematical proof showing that S does not include Ia. This is only significant because of the previous missunderstanding.

As the work group progressed on the rewriting, they reached a level of agreement on principles allowing work to begin on two other areas of need. These were to reconsider the hydrologic soils classifications recognizing the vastly expanded data base available today and the capabilities of modern computers, and to reconsider the tables of curve numbers in terms of the expanded rainfall-run-off database available.

## CONCEPT OF HYDROLOGIC SOIL GROUPS

There has been a vast increase in basic soils property data since Musgrave first proposed the concept of hydrologic soil groups in Handbook of Agriculture. The data are now available in an electronic database. Modern tools of data mining were explored for analysis of this mass of data. Both neural networks and fuzzy sets were tried with fuzzy sets being adopted.

Originally, Soil Hydrologic Groups were assigned to soil series and phase of series by soil scientists based upon their interpretation of the published criteria. The soil scientist=s interpretation of the published criteria has varied across time and between states or regions. Thus, the hydrologic group criteria are not applied consistently across the United States. This is most evident in the comparison of soils with similar soil hydrologic and physical properties and dissimilar hydrologic group placement.

The Hydrologic Soil Groups are A, B, C, D and dual groups A/D, B/D and C/D. Soils in hydrologic group A have low run-off potential. Soils that have a moderate rate of infiltration when thoroughly wet are in hydrologic group B. Hydrologic group C soils that have a slow rate of infiltration rate when thoroughly wet. Soils in hydrologic group D have a high run-off potential. Dual Hydrologic Soil Groups (A/D, B/D, and C/D) are given for certain wet soils that could be adequately drained. The first letter applies to the drained and the second to the undrained condition. Soils are assigned to dual groups if the shallow depth to a permanent water table is the sole criteria for assigning a soil to hydrologic group D.

A model or rule based automated system that provides for objective placement of soils into Hydrologic Soil Groups was developed. The fuzzy system model for assigning soils to hydrologic soil groups is based on the published hydrologic group assumptions and criteria.

The soil surface is taken to be bare and the soil is not permanently frozen. The soil physical and hydrologic characteristic which make up the hydrologic grouping criteria are the depth to permanent water, depth to a restrictive layer, minimum saturated hydraulic conductivity in the soil=s upper 100 cm, and the soil=s texture.

There are three components to the fuzzy systems model: the Property, the Evaluation, and the Rule. The Property is an SQL (Standard Query Language) statement that retrieves the needed soil data from the soil survey database. An example of a Property is the depth to a restrictive layer.

The Evaluation=s function is to apply the data received from the SQL statement to a statement of the property=s relevance to the soil=s hydrologic grouping. In the case of the depth to a restrictive layer, the Evaluation determines the fit or truthfulness of the statement, AThe run-off characteristics of the soil increases as a soil=s depth a restrictive layer becomes shallower.

At some depth, the restrictive layer in the soil has a maximum contribution to run-off and the Evaluation is true. The result of an Evaluation is some number between 0 and 1. This number represents the truthfulness of the statement being evaluated. The closer the number is to 1 the closer the soil=s property fits the grouping criterion. Conversely, the closer the number is to 0 the less the soil property=s contribution to the hydrologic grouping of soils. In the restrictive layer example, an Evaluation output of 1 would mean that the soil=s restrictive layer is shallower than 50cm. Any output less than 1 would mean that the depth to any soil restrictive layer is greater than 50cm. This numeric output from the evaluation is passed to the Rule. The Rule is the third component of the fuzzy system model. The Rule serves two functions that result in a soil=s Hydrologic Soil Group placement.

The first function is to provide tools for the construction and implementation of the grouping system=s model and to bring the various hydrologic grouping criterion evaluations together into a single Hydrologic Soil Group model. The second is to convert the model=s numeric output into a Hydrologic Soil Group. The model was applied to 1828 unique soil phases using data from Kansas, South Dakota, Missouri, Iowa, Wyoming, and Colorado and the correlation between these soils= assigned and modeled hydrologic grouping was analysed. Table shows a detailed comparison by Hydrologic Soil Group between the currently assigned HSG and the modeled HSG. The correlation between the assigned and modeled HSG A and HSG D soils is higher than the correlation between the assigned and modeled HSG B and HSG C soils.

There are several reasons for the poorer correlation between the assigned and modeled groups B and C. The first is that of the boundary condition which occurs when a soil has properties that do not fit entirely into a single hydrologic group. In this case, the soil scientist may have placed the soil into one HSG while the model placed the soil into an adjacent group. Groups B and C are the most prone to this error because they are bounded by two groups whereas HSG A and D are only bounded by one group. Another source of correlation inconsistency is that the assigned HSG may be relatively correct, but the data in the database may not support the corresponding HSG determination by the model. Finally, correlation inconsistencies can be attributed to the fuzzy modeling of the subjective Hydrologic Soil Group criteria.

**Table. Correlation Frequency Between Assigned and Fuzzy Modeled Hydrologic Soil Groups**

| Current HSG | Number of Soils | Fuzzy hsg Assignment Frequency | | | | | | |
|---|---|---|---|---|---|---|---|---|
| | | A | B | C | D | A/D | B/D | C/D |
| A | 155 | 0.9 | 0.08 | 0 | 0.01 | 0.01 | 0 | 0 |
| B | 821 | 0.25 | 0.54 | 0.17 | 0.02 | 0.01 | 0 | 0 |

| | | | | | | | | |
|---|---|---|---|---|---|---|---|---|
| C | 405 | 0.04 | 0.25 | 0.34 | 0.31 | 0 | 0.03 | 0.04 |
| D | 404 | 0.02 | 0.05 | 0.05 | 0.64 | 0.06 | 0.1 | 0.08 |
| A/D | 1 | 0 | 0 | 0 | 0 | 0 | 0.55 | 0 |
| B/D | 29 | 0.1 | 0.07 | 0.07 | 0 | 0.1 | 0.31 | 0.1 |
| C/D | 13 | 0 | 0.08 | 0.08 | 0.39 | 0 | | 0.15 |

## CURVE NUMBERS FROM RAINFALL-RUN-OFF DATA

The watershed research programme of the USDA, Agricultural Research Service is a continuation of research initiated by the Soil Conservation Service. Much of the data collected should be directly applicable to the determination of curve numbers and to explaining the variation of curve number with the soil-cover complex. In addition, Prof. R. H. Hawkins of the University of Arizona had been developing the world=s largest event rainfall–run-off data base with software to analyse the data. Prof. Hawkins was added to the ARS/NRCS work group.

The ideal method of determining curve numbers from observed data is elusive due to the stochastic nature of the variable. Our decision to emphasize the concept that the run-off equation serves to transform a rainfall frequency distribution into a run-off frequency distribution led to use of frequency matching. That is, curve numbers were determined by use of rainfall of a given return period with run-off of the same return period. These data may or may not come from the same storm. That is, frequency transformation leads to treating ordered pairs. In his analysis, Hawkins recognized that not all data sets are adequate to define a curve number and some watersheds do not even perform according to the Curve Number run-off equation.

He developed a graphical procedure in which the calculated curve number is plotted versus the precipitation used in calculating that curve number. In part, this plot is in recognition that, due to the random nature of the curve number, for a watershed with a given Atrue@ curve number, the actual event curve number will range above and below that Atrue@ value. For small rainfall events the event Ia will vary above and below the Atrue@ Ia. Curve numbers can only be determined if there is run-off, so if the event Ia is low, run-off will occur and a CN computed.

If the Ia is high, no run-off occurs so no CN can be computed. Thus, the process of computing CN for small events biases the CN towards high values (low Ia). The CN vs. P plot displays this bias and the storm magnitude at which the bias becomes insignificant.

The concept of the CN method being a transformation between a rainfall-depth distribution and a run-off depth distribution is applied in treating rainfall and run-off data. The rainfall depths and the run-off depths are sorted separately and then re–aligned on a rank order basis to form P:Q pairs of equal return period. The individual run-offs are not necessarily associated with the original causative rainfalls. When CNs are calculated from real storm data as outlined above, a secondary relationship almost always emerges between CN and storm rainfall depth itself. In most of these cases, these calculated CNs approach a constant value with increasing rainfall.Three variations on this theme have been observed, however, and are described in the following:

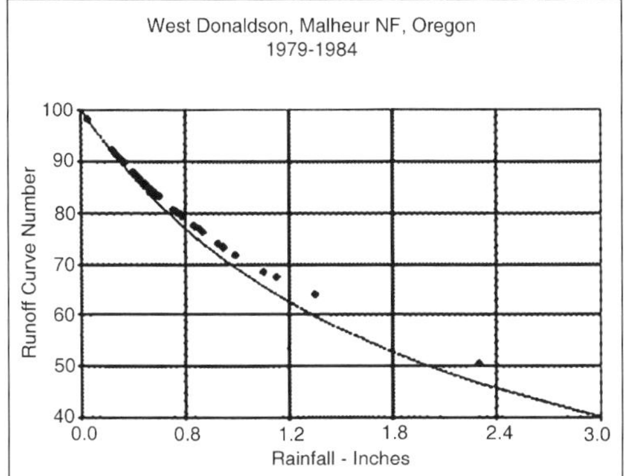


**Fig.** Complacent Behaviour

## Complacent Behaviour

Here the observed CN declines steadily with increasing rainfall depth, and with no appreciable tendency to achieve a stable value. An example of this is given in Figure above. Curve Numbers cannot be safely determined from data which exhibit this pattern, because no constant value is clearly approached. This Curve Number behaviour has been found to indicate a partial source area situation where the source area fraction may be quite small. In these cases the run-off is more properly modeled by the linear form Q=CP rather than by the Curve Number run-off equation.

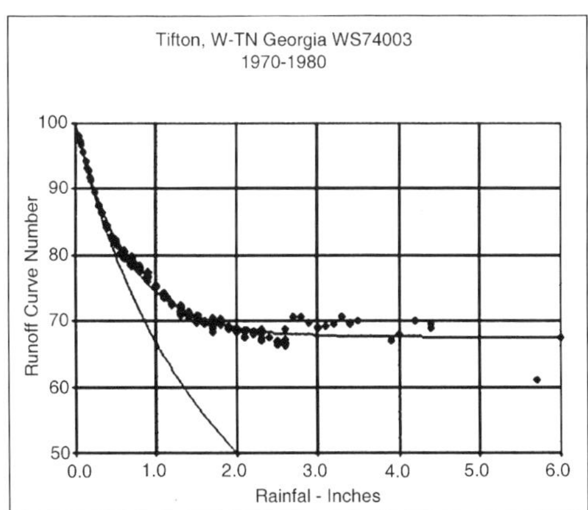


## Standard Behaviour

This is the most common scenario. The observed CN declines with increasing storm size, as in the Complacent situation described. However, here the CNs approach and/or maintain a near–constant value with increasingly larger storms. The run-off itself may arise from a variety of source processes, including overland flow and rapid subsurface flow. An example of this pattern is given in Figure.

**Violent Behaviour**

The distinguishing feature here is that the observed CNs rise suddenly and asymptotically approach an apparent constant value. There is often accompanying Complacent behaviour at lower rainfalls. From a source process standpoint, this could be a threshold phenomenon at some critical rainfall depth value. Rietz and Hawkins used their large electronic database of rainfall-run-off data to determine data-defined CNs calculated from local rainfall-run-off data. Their study attempted to develop a better understanding of a watershed=s land use as manifested in its CN. The land use variable was isolated in a large data set of small watersheds, and land use CN=s were calculated and analysed for each of these watersheds at a local, regional, and national scale.

Data in this study contained detailed land use information on 177 watersheds covering 2,455 years of record and 32,891 events. Watershed land uses analysed in this study were: alfalfa (closec–seed legume), corn (row crop), cotton (row crop), desert shrub, fallow, forest, grassland, meadow, oats (small grain), pasture, range, sage brush, sorghum (row crop), soy bean (row crop) and wheat (small grain). Many of the cultivated watersheds had a different land use year to year due to crop rotation. When one watershed had several land uses throughout the period of record, data were segregated by date of land use and CN calculated for each land use. Curve Numbers for each land use on each watershed were determined using the Asym-ptotic data-derived procedure with ordered P:Q pairs.

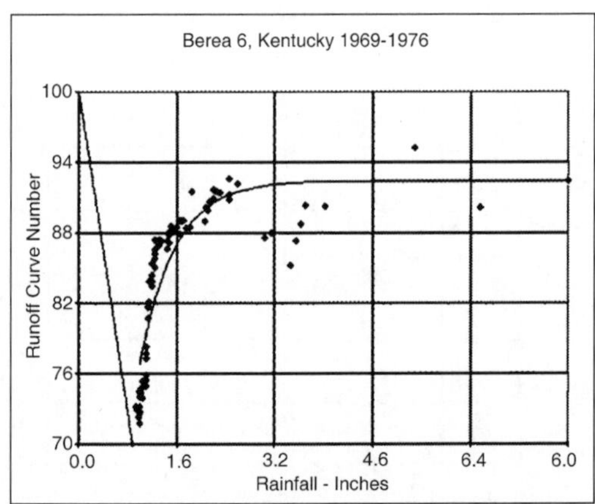


**Fig.** Violent Behaviour

Differences in land use CN at the local level can be attributed solely to a change in hydrological response of a watershed due to land use because all other input variables of the Curve Number model are constant. Soil type, climate, and morphology are fixed, and watershed land conditions are constant for that land use. Of the 177 watersheds used in this study, 53 were found to have more than one land use during their period of record.

When ANOVA independent group tests were performed 96.23 per cent of the watersheds (51 out of 53) were found to have significant differences among their land use Curve Numbers.

*The results were:*

- Meadow Curve Numbers were almost always the lowest CN for a watershed. Of the 19 watersheds that had a meadow in their crop rotation, 13 had meadow Curve Numbers significantly lower than any other land use CN within that same watershed. Three others had meadow Curve Numbers that ranked second lowest.
- Pasture is a grassland watershed that is grazed. This CN was generally significantly higher (6 out of 8 times) than all other land use Curve Numbers on a given watershed.
- Watersheds that experienced a land conversion had a significant difference in their data-derived CN after the conversion. Two desert watersheds in the Boco Mountains of Colorado were converted from desert brush to grass and with both, the grass Curve Number was significantly lower than desert brush. Another land use conversion study was at Ricsel, Texas watershed #42036. This ARS experimental watershed was 100 per cent rangeland that was Ainfested≅ with honey mesquite. In 1972 the watershed was treated, at which time the mesquites were killed and left standing. The Curve Number following mesquite killing was statistically higher than for live mesquites.

In order to compare land use Curve Numbers across all locations, a land use had to have been applied in more than one location, which was found to be the case with 11 single land uses. Seven of these 11 land uses (63.6 per cent) exhibited a significant difference.

The rank order of Curve Numbers was: forest and meadow were among the lowest average Curve Number, row crops or small grain were in mid-range, and desert brush exhibited the highest average Curve Number. Small grains (wheat and oats) were grouped together as were row crops (soybean, corn and sorghum). It was surprising however, that the average Curve Number for range was the third lowest average Curve Number and that row crops had a lower average CN than small grains.

*The results of the work group are:*

- The basic description of the Curve Number procedure is more clear, more consistent, and is in a more technically defensible state.
- The Hydrologic Soil Groups are associated with soil physical properties through fuzzy set procedures.

- Procedures for determining Curve Numbers from local data and interpretation of the results are much better established.

However, more work is needed. For example, we still need to work at verifying, adjusting, and correcting the table of CN. Regional variation in Curve Numbers should be explored. Finally, the whole issue of the application of Curve Numbers in continuous simulation models (CREAMS, GLEAMS, EPIC, SWAT) may be quite different from the design storm Curve Numbers of NEH–4. If there is a relation, that relation should be determined.

## WATER PROBLEM IN INDIA

Summers are here and the cities in India are already complaining about water shortage not to mention many villages which lack safe drinking water. In the list of 122 countries rated on quality of portable water, India ranks a lowly 120.

Although India has 4 per cent of the world's water, studies show average availability is shrinking steadily. It is estimated that by 2020, India will become a water-stressed nation. Nearly 50 per cent of villages still don't have any source of protected drinking water. According to 2001 census 68.2 per cent households have access to safe drinking water. The department of drinking water supply estimates that 94 per cent of rural habitations and 91 per cent urban households have access to drinking water. But according to experts these figures are misleading simply because coverage refers to installed capacity and not actual supply.

The ground reality is that of the 1.42 million villages in India, 1, 95,813 are affected by chemical contamination of water. The quality of ground water which accounts of more than 85 per cent of domestic supply is a major problem in many areas as none of the rivers have water fit to drink. 37.7 million People – over 75 per cent of whom are children are afflicted by waterborne diseases every year.

Overdependence on groundwater has brought in contaminants, fluoride being one of them. Nearly 66 million people in 20 states are at risk because of the excessive fluoride in water. While the permissible limit of fluoride in water is 1 mg per litre in states like Haryana it is as high as 48 mg in some places. Delhi water too has 32 mg. But the worst hits are Rajasthan, Gujarat and Andhra Pradesh. Nearly 6 million children below 14 suffer from dental, skeletal and non-skeletal fluorosis.

Arsenic is the other big killer lurking in ground water putting at risk nearly 10 million people. The problem is acute in Murshidabad, Nadia, North and South 24 Paraganas, Malda and Vardhaman districts of West Bengal. The deeper aquifers in the entire Gangetic plains contain arsenic.

High nitrate content in water is another serious concern. Fertilizers, septic tanks, sewage tanks etc. are the main sources of nitrate contamination. The groundwater in MP, UP, Punjab, Haryana, Delhi, Kanataka and Tamil Nadu has shown traces of nitrates.

However it is bacteriological contamination which leads to diarrhoea, cholera and hepatitis which is widespread in India. A bacteriological analysis of the water in Bangalore revealed 75 per cent bore wells were contaminated. Iron; hardness and salinity are also a concern. Nearly 12,500 habitats have been affected by salinity.

In Gujarat it is a major problem in coastel districts. Often babies die of dehydration and there are major fights in villages for freshwater. Some villages have seen 80 per cent migration due to high salinity.

Health is not the only issue; impure water is a major burden on the state as well. Till the 10th plan the government had spent ₹ 1,105 billion on drinking water schemes. Yet it is the poor who pay a heavier price spending around ₹ 6700 crore annually on treatment of waterborne diseases.

There is an urgent need to look for alternative sources of portable water in places where water quality has deteriorated sharply. Community based water quality monitoring guidelines should be encouraged. People should be encouraged to look at traditional methods of protecting water sources. Also in places where groundwater has arsenic or fluoride, surface water should be considered as an alternative.

## WATER SUPPLY AND SANITATION IN INDIA

Water supply and sanitation in India continue to be inadequate, despite longstanding efforts by the various levels of government and communities at improving coverage. The level of investment in water and sanitation, albeit low by international standards, has increased during the 2000s. Access has also increased significantly.

For example, in 1980 rural sanitation coverage was estimated at 1 per cent and reached 21 per cent in 2008. Also, the share of Indians with access to improved sources of water has increased significantly from 72 per cent in 1990 to 88 per cent in 2008.

At the same time, local government institutions in charge of operating and maintaining the infrastructure are seen as weak and lack the financial resources to carry out their functions. In addition, no major city in India is known to have a continuous water supply and an estimated 72 per cent of Indians still lack access to improved sanitation facilities.

A number of innovative approaches to improve water supply and sanitation have been tested in India, in particular in the early 2000s. These include demand-driven approaches in rural water supply since 1999, community-led total sanitation, a public-private partnerships to improve the continuity of urban water supply in Karnataka, and the use of micro-credit to women in order to improve access to water.

## ACCESS

In 2008, 88 per cent of the population in India had access to an improved water source, but only 31 per cent had access to improved sanitation. In rural areas, where 72 per cent of India's population lives, the respective shares are 84 per cent for water and only 21 per cent for sanitation. In urban areas, 96 per cent had access to an improved water source and 54 per cent to improved sanitation. Access has improved substantially since 1990 when it was estimated to stand at 72 per cent for water and 18 per cent for sanitation.

As of 2010, the UN estimated based on Indian statistics that 626 people practice open defecation. In June 2012 Minister of Rural Development Jairam Ramesh stated India is the worlds largest "open air toilet". He also remarked that Pakistan, Bangladesh and Afghanistan have better sanitation records.

According to Indian norms, access to improved water supply exists if at least 40 litres/capita/day of safe drinking water are provided within a distance of 1.6 km or 100 meter of elevation difference, to be relaxed as per field conditions. There should be at least one pump per 250 persons.

## SERVICE QUALITY

Water and sanitation service quality in India is generally poor, although there has been some limited progress concerning continuity of supply in urban areas and access to sanitation in rural areas.

## WATER SUPPLY

Challenges. None of the 35 Indian cities with a population of more than one million distribute water for more than a few hours per day, despite generally sufficient infrastructure. Owing to inadequate pressure people struggle to collect water even when it is available. According to the World Bank, none have performance indicators that compare with average international standards. A 2007 study by the Asian Development Bank showed that in 20 cities the average duration of supply was only 4.3 hours per day. No city had continuous supply. The longest duration of supply was 12 hours per day in Chandigarh, and the lowest was 0.3 hours per day in Rajkot. In Delhi residents receive water only a few hours per day because of inadequate management of the distribution system. This results in contaminated water and forces households to complement a deficient public water service at prohibitive 'coping' costs; the poor suffer most from this situation. For example, according to a 1996 survey households in Delhi spent an average of 2,182 (US$39.5) per year in time and money to cope with poor service levels. This is more than three times as much as the 2001 water bill of about US$18 per year of a Delhi household that uses 20 cubic meters per month.

Achievements. Jamshedpur, a city in Jharkhand with 573,000 inhabitants, provided 25 per cent of its residents with continuous water supply in 2009.

Navi Mumbai, a planned city with more than 1m inhabitants, has achieved continuous supply for about half its population as of January 2009. Badlapur, another city in the Mumbai Conurbation with a population of 140,000, has achieved continuous supply in 3 out of 10 operating zones, covering 30 per cent of its population. Thiruvananthapuram, the capital of Kerala state with a population of 745,000 in 2001, is probably the largest Indian city that enjoys continuous water supply.

## SANITATION

Most Indians depend on on-site sanitation facilities. Recently, access to on-site sanitation have increased in both rural and urban areas. In rural areas, total sanitation has been successful. In urban areas, a good practice is the Slum Sanitation Programme in Mumbai that has provided access to sanitation for a quarter million slum dwellers.

Sewerage, where available, is often in a bad state. In Delhi the sewerage network has lacked maintenance over the years and overflow of raw sewage in open drains is common, due to blockage, settlements and inadequate pumping capacities. The capacity of the 17 existing wastewater treatment plants in Delhi is adequate to cater a daily production of waste water of less than 50 per cent of the drinking water produced. Of the 2.5 Billion people in the world that defecate openly, some 665 million live in India. This is of greater concern as 88 per cent of deaths from diarrhoea occur because of unsafe water, inadequate sanitation and poor hygiene.

## ENVIRONMENT

As of 2003, it was estimated that only 27 per cent of India's wastewater was being treated, with the remainder flowing into rivers, canals, groundwater or the sea., For example, the sacred Ganges river is infested with diseases and in some places "the Ganges becomes black and septic. Corpses, of semi-cremated adults or enshrouded babies, drift slowly by.". News Week describes Delhi's sacred Yamuna River as "a putrid ribbon of black sludge" where fecal bacteria is 10,000 over safety limits despite a 15-year programme to address the problem. Cholera epidemics are not unknown.

## HEALTH IMPACT

The lack of adequate sanitation and safe water has significant negative health impacts including diarrhoea, referred to by travellers as the "Delhi Belly", and experienced by about 10 million visitors annually. While most visitors to India recover quickly and otherwise receive proper care, the World Health Organisation estimated that around 700,000 Indians die each year from diarrhoea. The dismal working conditions of sewer workers are another concern. A survey of the working conditions of sewage workers in Delhi showed that most of them suffer from chronic diseases, respiratory problems, skin disorders, allergies, headaches and eye infections.

## WATER SUPPLY AND WATER RESOURCES

Depleting ground water table and deteriorating ground water quality are threatening the sustainability of both urban and rural water supply in many parts of India. The supply of cities that depend on surface water is threatened by pollution, increasing water scarcity and conflicts among users. For example, Bangalore depends to a large extent on water pumped since 1974 from the Kaveri river, whose waters are disputed between the states of Karnataka and Tamil Nadu. As in other Indian cities, the response to water scarcity is to transfer more water over large distances at high costs. In the case of Bangalore, the 3,384 crore (US$612.5 million) Kaveri Stage IV project, Phase II, includes the supply of 500,000 cubic meter of water per day over a distance of 100 km, thus increasing the city's supply by two thirds.

## RESPONSIBILITY FOR WATER SUPPLY AND SANITATION

Water supply and sanitation is a State responsibility under the Indian Constitution. States may give the responsibility to the Panchayati Raj Institutions (PRI) in rural areas or municipalities in urban areas, called Urban Local Bodies (ULB). At present, states generally plan, design and execute water supply schemes (and often operate them) through their State Departments (of Public Health Engineering or Rural Development Engineering) or State Water Boards.

Highly centralised decision-making and approvals at the state level, which are characteristic of the Indian civil service, affect the management of water supply and sanitation services. For example, according to the World Bank in the state of Punjab the process of approving designs is centralised with even minor technical approvals reaching the office of chief engineers. A majority of decisions are made in a very centralised manner at the headquarters. In 1993 the Indian constitution and relevant state legislations were amended in order to decentralise certain responsibilities, including water supply and sanitation, to municipalities. Since the assignment of responsibilities to municipalities is a state responsibility, different states have followed different approaches. According to a Planning Commission report of 2003 there is a trend to decentralise capital investment to engineering departments at the district level and operation and maintenance to district and gram panchayat levels.

## POLICY AND REGULATION

The responsibility for water supply and sanitation at the central and state level is shared by various Ministries. At the central level, The Ministry of Rural Development is responsible for rural water supply through its Department of

Drinking Water Supply (DDWS) and theMinistry of Housing and Urban Poverty Alleviation is responsible for urban water supply. However, except for the National Capital Territory of Delhi and other Union Territories, the central Ministries only have an advisory capacity and a very limited role in funding. Sector policy thus is a prerogative of state governments.

## SERVICE PROVISION

Urban areas. Institutional arrangements for water supply and sanitation in Indian cities vary greatly. Typically, a state-level agency is in charge of planning and investment, while the local government (Urban Local Bodies) is in charge of operation and maintenance. Some of the largest cities have created municipal water and sanitation utilities that are legally and financially separated from the local government. However, these utilities remain weak in terms of financial capacity. In spite of decentralisation, ULBs remain dependent on capital subsidies from state governments. Tariffs are also set by state governments, which often even subsidise operating costs. Furthermore, when no separate utility exists there is no separation of accounts for different activities within a municipality. Some states and cities have non-typical institutional arrangements. For example, in Rajasthan the sector is more centralised and the state government is also in charge of operation and maintenance, while in Mumbai the sector is more decentralised and local government is also in charge of planning and investment.

Private sector participation. The private sector plays a limited, albeit recently increasing role in operating and maintaining urban water systems on behalf of ULBs. For example, the Jamshedpur Utilities and Services Company (Jusco), a subsidiary of Tata Steel, has a lease contract for Jamshedpur (Jharkhand), a management contract in Haldia (West Bengal), another contract in Mysore (Karnataka) and since 2007 a contract for the reduction of non-revenue water in parts of Bhopal (Madhya Pradesh). The French water company Veolia won a management contract in three cities in Karnataka in 2005.

In 2002 a consortium including Thames Water won a pilot contract covering 40,000 households to reduce non-revenue water in parts of Bangalore, funded by the Japan Bank for International Cooperation. The contract was scaled up in 2004. The Cypriot company Hydro-Comp, together with two Indian companies, won a 10-year concession contract for the city of Latur City (Maharashtra) in 2007 and an operator-consultant contract in Madurai (Tamil Nadu). Furthermore, the private Indian infrastructure development company SPML is engaged in Build-Operate-Transfer (BOT) projects, such as a bulk water supply project for Bhiwandi (Maharashtra).

Rural areas. There are about a 100,000 rural water supply systems in India. At least in some states responsibility for service provision is in the process of being partially transferred from State Water Boards and district governments to Panchayati Raj Institutions (PRI) at the block or village level (there were about 604 districts and 256,000 villages in India in 2002, according to Subdivisions of India. Blocks are an intermediate level between districts and villages). Where this transfer has been initiated, it seems to be more advanced for single-village water schemes than for more complex multi-village water schemes. Despite their professed role Panchayati Raj Institutions, play only a limited role in provision of rural water supply and sanitation as of 2006. There has been limited success in implementing decentralisation, partly due to low priority by some state governments. Rural sanitation is typically provided by households themselves in the form of latrines.

## COMMUNITY-LED TOTAL SANITATION

In 1999 a demand-driven and people-centered sanitation programme was initiated under the name Total Sanitation Campaign (TSC) or Community-led total sanitation. It evolved from the limited achievements of the first structured programme for rural sanitation in India, the Central Rural Sanitation Programme, which had minimal community participation. The main goal of Total Sanitation Campaign is to eradicate the practice of open defecation by 2017. Community-led total sanitation is not focused on building infrastructure, but on preventing open defecation through peer pressure and shame. In Maharashtra where the programme started more than 2000 Gram Panchayats have achieved "open defecation free" status. Villages that achieve this status receive monetary rewards and high publicity under a programme called Nirmal Gram Puraskar.

## DEMAND-DRIVEN APPROACHES IN RURAL WATER SUPPLY

Most rural water supply schemes in India use a centralised, supply-driven approach, *i.e.* a government institution designs a project and has it built with little community consultation and no capacity building for the community, often requiring no water fees to be paid for its subsequent operation. Since 2002 the Government of India has rolled out at the national level a programme to change the way in which water and sanitation services are supported in rural areas. The programme, called *Swajaldhara*, decentralises service delivery responsibility to rural local governments and user groups. Under the new approach communities are being consulted and trained, and users agree up-front to pay a tariff that is set at a level sufficiently high to cover operation and maintenance costs. It also includes measures to promote sanitation and to improve hygiene behaviour. The national programme follows a pilot programme launched in 1999.

According to a 2008 World Bank study in 10 Indian states, *Swajaldhara* results in lower capital costs, lower administrative costs and better service quality compared to the supply-driven approach. In particular, the study found that the average full cost of supply-driven schemes is ₹ 38 (US$0.7) per cubic meter, while it is only ₹ 26 (US$0.5) per cubic meter for demand-driven schemes. These costs include capital, operation and maintenance costs, administrative costs and coping costs incurred by users of malfunctioning systems. Coping costs include travelling long distances to obtain water, standing in long queues, storing water and repairing failed systems. Among the surveyed systems that were built using supply-driven approach system breakdowns were common, the quantity and quality of water supply were less than foreseen in designs, and 30 per cent of households did not get daily supply in summer. The poor functioning of one system sometimes leads to the construction of another system, so that about 30 per cent of households surveyed were served by several systems. As of 2008 only about 10 per cent of rural water schemes built in India used a demand-driven approach. Since water users have to pay lower or no tariffs under the supply-driven approach, this discourages them to opt for a demand-driven approach, even if the likelihood of the systems operating on a sustainable basis is higher under a demand-driven approach.

## ACHIEVING CONTINUOUS WATER SUPPLY

In the cities of Hubli, Belgaum and Gulbarga in the state of Karnataka, the private operator Veolia increased water supply from once every 2–15 days for 1–2 hours, to 24 hours per day for 180,000 people (12 per cent of the population of the 3 cities) within 2 years (2006–2008). This was achieved by carefully selecting and ring-fencing demonstration zones (one in each city), renovating the distribution network, installing meters, introducing a well-functioning commercial system, and effective grass-roots social intermediation by an NGO, all without increasing the amount of bulk water supplied. The project, known by its acronym as KUWASIP (Karnataka Urban Water Sector Improvement Project), was supported by a US$39.5 million loan from the World Bank. It constitutes a milestone for India, where no large city so far has achieved continuous water supply. The project is expected to be scaled-up to cover the entire area of the three cities.

## MICRO-CREDIT FOR WATER CONNECTIONS IN TAMIL NADU

In Tiruchirapalli in Tamil Nadu, the NGO Gramalaya, established in 1987, and women self-help groups promote access to water supply and sanitation by the poor through micro-credit. Among the benefits are that women can spend more time with their children, earn additional income, and sell surplus water to neighbours. This money contributes to her repayment of the WaterCredit loan. The initiative is supported by the US-based non-profit Water Partners International.

## THE JAMSHEDPUR UTILITIES AND SERVICES COMPANY

The Jamshedpur Utilities and Services Company (JUSCO) provides water and sanitation services in Jamshedpur, a major industrial centre in East India that is home to Tata Steel. Until 2004 a division of Tata Steel provided water to the city's residents. However, service quality was poor with intermittent supply, high water losses and no metering. To improve this situation and to establish good practices that could be replicated in other Indian cities, JUSCO was set up as a wholly owned subsidiary of Tata Steel in 2004.

Efficiency and service quality improved substantially over the following years. The level on non-revenue water decreased from an estimated 36 per cent in 2005 to 10 per cent in 2009; one quarter of residents received continuous water supply (although the average supply remained at only 7 hours per day) in 2009; the share of metered connections increased from 2 per cent in 2007 to 26 per cent in 2009; the number of customers increased; and the company recovered its operating costs plus a portion of capital costs. Identifying and legalising illegal connections was an important element in the reduction of non-revenue water.

The utility prides itself today of the good drinking water quality provided and encourages its customers to drink from the tap. The utility also operates a wastewater treatment plant that meets discharge standards. The private utility pays salaries that are higher than civil service salaries and conducts extensive training programmes for its staff. It has also installed a modern system to track and resolve customer complaints. Furthermore, it conducts independent annual customer satisfaction surveys. JUSCO's vision is to be the preferred provider of water supply and other urban services throughout India. Together with Ranhill Malaysia it won a 25-year concession contract for providing the water supply in Haldia City, West Bengal.

## EFFICIENCY

There are only limited data on the operating efficiency of utilities in India, and even fewer data on the efficiency of investments.

Concerning operating efficiency, a study of 20 cities by the Jawaharlal Nehru National Urban Renewal Mission with the support of the Asian Development Bank showed an average level of non-revenue water (NRW) of 32 per cent. However, 5 out of the 20 cities did not provide any data. For those that provided data there probably is a large margin of error, since only 25 per cent of connections are metered, which makes it very difficult to estimate non-revenue water.

Also, three utilities show NRW levels of less than 20 per cent, two of which have practically no metering, which indicates that the numbers are not reliable and actual values are likely to be higher. In Delhi, which was not included in the ADB study, non-revenue water stood at 53 per cent and there were about 20 employees per 1000 connections. Furthermore, only 70 per cent of revenue billed was actually collected.

Concerning labour productivity, the 20 utilities in the sample had on average 7.4 employees per 1,000 connections, which is much higher than the estimated level for an efficient utility. A survey of a larger sample of Indian utilities showed an average ratio of 10.9 employees per 1,000 connections.

## TARIFFS, COST RECOVERY AND SUBSIDIES

Water and sewer tariffs in India are low in both urban and rural areas. In urban areas they were set at the equivalent of about US$0.10 per cubic meter in 2007 and recovered about 60 per cent of operating and maintenance costs, with large differences between cities. Some cities such as Kolkata do not bill residential users at all.

In rural areas the level of cost recovery often is even lower than in urban areas and was estimated at only 20 per cent in rural Punjab. Subsidies were estimated at US$1.1 billion per year in the mid-1990s, accounting to 4 per cent of all government subsidies in India. 70 per cent of those benefiting from the subsidies are not poor.

## URBAN AREAS

Metering. Water metering is the precondition for billing water users on the basis of volume consumed. According to a 1999 survey of 300 cities about 62 per cent of urban water customers in metropolitan areas and 50 per cent in smaller cities are metered (average 55 per cent). However, meters often do not work so that many "metered" customers are charged flat rates. Bangalore and Pune are among the few Indian cities that meter all their customers. Many other cities have no metering at all or meter only commercial customers. Users of standposts receive water free of charge. A 2007 study of 20 cities by the Jawaharlal Nehru National Urban Renewal Mission with the support of the Asian Development Bank (ADB) showed that only 25 per cent of customers of these utilities were metered. Most other customers paid a flat tariff independent of consumption. Some utilities, such as the one serving Kolkata, actually do not bill residential users at all.

Tariff levels. According to the same ADB study the average tariff for all customers – including industrial, commercial and public customers – is ₹ 4.9 (US$0.1) per cubic meter. According to a 2007 global water tariff survey by the OECD the residential water tariff for a consumption of 15 $m^3$ was equivalent to US$0.15 per $m^3$ in Bangalore, US$0.12 per $m^3$ in Calcutta, US$0.11 per m3 in New Delhi and US$0.09 per $m^3$ in Mumbai. Only Bangalore had a sewer tariff of US$0.02 per $m^3$. The other three cities did not charge for sewerage, although the better-off tend to be the ones with access to sewers.

Tariff structure. The tariff for customers that are effectively metered is typically a uniform linear tariff, although some cities apply increasing-block tariffs.

Affordability. Urban water tariffs were highly affordable according to data from the year 2000. A family of five living on the poverty line which uses 20 cubic meter of water per month would spend less than 1.2 per cent of its budget on its water bill if it had a water meter. If it did not have a water meter and was charged a flat rate, it would pay 2.0 per cent of its budget. This percentage lies below the often used affordability threshold of 5 per cent. However, at that time the average metered tariff was estimated at only US$0.03 per $m^3$, or less than three times what it was estimated to be in 2007. Apparently no more up-to-date estimates on the share of the average water bill in the budget of the poor are available.

Cost recovery. According to a 2007 study of 20 cities the average rate of cost recovery for operating and maintenance costs of utilities in these cities was 60 per cent. Seven of the 20 utilities generated a cash surplus to partially finance investments. Chennai generated the highest relative surplus. The lowest cost recovery ratio was found in Indore in Madhya Pradesh, which recovered less than 20 per cent of its operating and maintenance costs.

Delhi example. In Delhi revenues were just sufficient to cover about 60 per cent of operating costs of the city's utility in 2004; maintenance has, as a result, been minimal. In the past, the Delhi utility has relied heavily on government financial support for recurrent and capital expenditures in the magnitude of ₹ 3 billion (US$54.3 million) per year and ₹ 7 billion (US$126.7 million) respectively.

As financial support for both capital and recurrent expenditures has been passed on as loans by the Government of the National Capital Territory of Delhi, the utility's balance sheet is loaded with a huge debt totalling about ₹ 50 billion (US$905 million) that it is unlikely to be able to service. Accounts receivable represent more than 12 months of billing, part of it being non recoverable.

The average tariff was estimated atUS$0.074/$m^3$ in 2001, compared to production costs of US$0.085/$m^3$, the latter probably being a very conservative estimate that does not take into account capital costs. Challenges faced in attempting to increase tariffs.

Even if users are willing to pay more for better services, political interests often prevent tariffs from being increased even to a small extent. An example is the city of Jabalpur where the central government and the state government financed a ₹ 130 million (US$2.4 million) water supply project from 2000–2004 to be operated by the Jabalpur Municipal Corporation, an entity that collected only less than half of its operational costs in revenues even before this major investment.

Even so the municipal corporation initially refused to increase tariffs. Only following pressure from the state government it reluctantly agreed to increase commercial tariffs, but not residential tariffs.

## RURAL AREAS

Cost recovery in rural areas is low and a majority of the rural water systems are defunct for lack of maintenance. Some state governments subsidise rural water systems, but funds are scarce and insufficient. In rural areas in Punjab, operation and maintenance cost recovery is only about 20 per cent. On one hand, expenditures are high due to high salary levels, high power tariff and a high number of operating staff. On the other hand, revenue is paid only by the 10 per cent of the households who have private connections. Those drawing water from public stand posts do not pay any water charges at all, although the official tariff for public stand post users is ₹ 15 (US$0.3) per month per household.

## SUBSIDIES AND TARGETING OF SUBSIDIES

There are no accurate recent estimates of the level of subsidies for water and sanitation in India. It has been estimated that transfers to the water sector in India amounted to ₹ 5,470.8 crore (US$990.2 million) per year in the mid-1990s, accounting for 4 per cent of all government subsidies in India. About 98 per cent of this subsidy is said to come from State rather than Central budgets. This figure may only cover recurrent cost subsidies and not investment subsidies, which are even higher. There is little targeting of subsidies. According to the World Bank, 70 per cent of those benefiting from subsidies for public water supply are not poor, while 40 per cent of the poor are excluded because they do not have access to public water services.

## INVESTMENT AND FINANCING

Investment in urban water supply and sanitation has increased during the first decade of the 21st century, not least thanks to increased central government grants made available under Jawaharlal Nehru National Urban Renewal Mission alongside with loans from the Housing and Urban Development Corporation.

## INVESTMENT

The Eleventh Five-Year Plan (2007–2012) foresees investments of ₹ 127,025 crore (US$23 billion) for urban water supply and sanitation, including urban (stormwater) drainage and solid waste management.

## FINANCING

55 per cent of the investments foreseen under the 11th Plan are to be financed by the central government, 28 per cent by state governments, 8 per cent by "institutional financing" such as HUDCO, 8 per cent by external agencies and 1.5 per cent by the private sector. Local governments are not expected to contribute to the investments. The volume of investments is expected to double

to reach 0.7 per cent of GDP. Also, it implies a shift in financing from state governments to the central government. During the 9th Plan only 24 per cent of investments were financed by the central government and 76 per cent by state governments. Central government financing was heavily focused on water supply in rural areas.

## INSTITUTIONS

State Financing Corporations (SFC) play an important role in making recommendations regarding the allocation of state tax revenues between states and municipalities, criteria for grants, and measures to improve the financial position of municipalities. According to the Planning Commission, SFCs are in some cases not sufficiently transparent and/or competent, have high transactions costs, and their recommendations are sometimes not being implemented. An important source of financing are loans from Housing and Urban Development Corporation Ltd (HUDCO), a Central government financial undertaking. HUDCO loans to municipal corporations need to be guaranteed by state governments.

HUDCO also on-lends loans from foreign aid, including Japanese aid, to states. The Jawaharlal Nehru National Urban Renewal Mission initiated in 2005 also plays an increasingly important role in financing urban water supply and sanitation through central government grants. The current system of financing water supply and sanitation is fragmented through a number of different national and state programmes. This results in simultaneous implementation with different and conflicting rules in neighbouring areas. In rural areas different programmes undermine each other, adversely affecting demand driven approaches requiring cost sharing by users.

## EXTERNAL COOPERATION

In absolute terms India receives almost twice as much development assistance for water, sanitation and water resources management as any other country, according to data from the Organisation for Economic Co-operation and Development. India accounts for 13 per cent of commitments in global water aid for 2006–07, receiving an annual average of about US$830 million (□620 million), more than double the amount provided to China. India's biggest water and sanitation donor is Japan, which provided US$635 million, followed by the World Bank with US$130 million. The annual average for 2004–06, however, was about half as much at US$448 million, of which Japan provided US$293 million and the World Bank US$87 million. The Asian Development Bank and Germany are other important external partners in water supply and sanitation.

In 2003 the Indian government decided it would only accept bilateral aid from five countries (the United Kingdom, the United States, Russia, Germany and Japan). A further 22 bilateral donors were asked to channel aid through non-governmental organisations, United Nations agencies or multilateral institutions such as the European Union, the Asian Development Bank or the World Bank.

## ASIAN DEVELOPMENT BANK

India has increased its loans from the Asian Development Bank (ADB) since 2005 after the introduction of new financing modalities, such as the multitranche financing facility (MFF) which features a framework agreement with the national government under which financing is provided in flexible tranches for subprojects that meet established selection criteria. In 2008 four MFFs for urban development investment programmes were under way in North Karnataka (US$862 million), Jammu and Kashmir (US$1,260 million),Rajasthan (US$450 million), and Uttarakhand (US$1,589 million). Included in these MFFs are major investments for the development of urban water supply and sanitation services.

## GERMANY

Germany supports access to water and sanitation in India through financial cooperation by KfW development bank and technical cooperation by GTZ. Since the early 1990s both institutions have supported watershed management in rural Maharashtra, using a participatory approach first piloted by the Social Centre in Ahmednagar and that constituted a fundamental break with the previous top-down, technical approach to watershed management that had yielded little results. The involvement of women in decision-making is an essential part of the project. While the benefits are mostly in terms of increased agricultural production, the project also increases availability of water resources for rural water supply. In addition, GTZ actively supports the introduction of ecological sanitationconcepts in India, including community toilets and decentralised wastewater systems for schools as well as small and medium enterprises. Many of these systems produce biogas from wastewater, provide fertilizer and irrigation water.

## JAPAN

As India's largest donor in the sector the Japan International Cooperation Agency (JICA) finances a multitude of projects with a focus on capital-intensive urban water supply and sanitation projects, often involving follow-up projects in the same locations.

Current projects. Projects approved between 2006 and 2009 include the Guwahati Water Supply Project (Phases I and II) in Assam, the Kerala Water Supply Project (Phased II and III), the Hogenakkal Water Supply and Fluorosis Mitigation Project (Phases I and II) in Tamil Nadu, the Goa Water Supply and Sewerage Project, the Agra Water Supply Project, the Amritsar Sewerage

Project in Punjab, the Orissa Integrated Sanitation Improvement Project, and the Bangalore Water Supply and Sewerage Project (Phase II).

Evaluation of past projects. An ex-post evaluation of one large programme, the Urban Water Supply and Sanitation Improvement Programme, showed that "some 60 per cent–70 per cent of the goals were achieved" and that "results were moderate". The programme was implemented by the Housing and Urban Development Corporation, Ltd. (HUDCO) from 1996 to 2003 in 26 cities. The evaluation says that "state government plans were not based on sufficient demand research, including the research for residents' willingness to pay for services", so that demand for connections was overestimated. Also fees (water tariffs) were rarely increased despite recommendations to increase them. The evaluation concludes that "HUDCO was not able to make significant contributions to the effectiveness, sustainability, or overall quality of individual projects. One of the reasons that not much attention was given to this problem is probably that there was little risk of default on the loans thanks to state government guarantees."

## WORLD BANK

Current projects. The World Bank finances a number of projects in urban and rural areas that are fully or partly dedicated to water supply and sanitation. In urban areas the World Bank supports the Andhra Pradesh Municipal Development Project (approved in 2009, US$300 million loan), the Karnataka Municipal Reform Project (approved in 2006, US$216 million loan), the Third Tamil Nadu Urban Development Project (approved in 2005, US$300 million loan) and the Karnataka Urban Water Sector Improvement Project (approved in 2004, US$39.5 million loan).

In rural areas it supports the Andhra Pradesh Rural Water Supply and Sanitation (US$150 million loan, approved in 2009), the Second Karnataka Rural Water Supply and Sanitation Project (approved in 2001,US$151.6 million loan), the Uttaranchal Rural Water Supply and Sanitation Project (approved in 2006, US$120 million loan) and the Punjab Rural Water Supply and Sanitation Project (approved in 2006, US$154 million loan).

Evaluation of past projects. A study by the World Bank's independent evaluation department evaluated the impact of the World Bank-supported interventions in the provision of urban water supply and wastewater services in Mumbai between 1973 and 1990.

It concluded that water supply and sewerage planning, construction and operations in Bombay posed daunting challenges to those who planned and implemented the investment programme. At the outset, there was a huge backlog of unmet demand because of under investment. Population and economic growth accelerated in the following decades and the proportion of the poor increased as did the slums which they occupied. The intended impacts of the programme have not been realised. Shortcomings include that "water is not safe to drink; water service, especially to the poor, is difficult to access and is provided at inconvenient hours of the day; industrial water needs are not fully met; sanitary facilities are too few in number and often unusable; and urban drains, creeks and coastal waters are polluted with sanitary and industrial wastes."

## WATER CRISIS LOOMS COUNTRYWIDE

Water has become the most commercial products of the century. This may sound bizarre, but true. In fact, what water is to the 21st century, oil was to the 20th century. The stress on the multiple water resources is a result of a multitude of factors. On the one hand, the rapidly rising population and changing lifestyles have increased the need for fresh water. On the other hand, intense competitions among users in agriculture, industry and domestic sector is pushing the ground water table deeper.

To get bucket of drinking water is a struggle for most women in the country. The virtually dry and dead water resources have lead to acute water scarcity, affecting the socio-economic condition of the society. The drought conditions have pushed villagers to move to cities in search of jobs, whereas women and girls have to trudge further. This time lost in fetching water can very well translate into financial gains, leading to a better life for the family. If opportunity costs were taken into account, it would be clear that in most rural areas, households are paying far more for water supply than the often-normal rates charged in urban areas. Also, if this cost of fetching water which is almost equivalent. to 150 million women days each year, is covered into a loss for the national exchequer, it translates into a whopping 10 billion rupees per year.

The government has accorded the highest priority to rural drinking water for ensuring universal access as a part of policy framework to achieve the goal of reaching the unreached. Despite the installation of more than 3.5 million hand pumps and over 116 thousand piped water supply schemes, in many parts of the country, the people face water scarcity almost every year, there by meaning that our water supply systems are failing to sustain despite huge investments.

In India, there are many villages either with scarce water supply or without any source of water. If there is no source of potable water in 2.5 kilometres, then the village becomes no source water village or problem village. In many rural areas, women still have to walk a distance of about 2.5 kms to reach the source of water. She reaches home carrying heavy pots, not to rest but to do other household chores of cooking, washing~ cleaning, caring of children and looking after livestock. Again in the evening she has to fetch water. Thus a rural woman's life is sheer drudgery. Water is the biggest crisis facing India in terms of spread and severity, affecting one in every three persons. Even in

Chennai, Bangalore, Shimla and Delhi, water is being rationed and India's food security is under threat. With the lives and livelihood of millions at risk, urban India is screaming for water. For instance, water is rationed twice a week in Bangalore, and for 30 minutes a day in Bhopal; 250 tankers make 2,250 trips to quench Chennai's thirst. Mumbai routinely lives through water cuts from January to June, when some areas get water once in three days in Hyderabad.

Harrowing midnight for a precious bucket of drinking water is a regular feature for many families of Vypuri, an Island off the mainland of Kochi. Women have to queue up in front of the public water taps, being at the lag end of the pipeline system, they get water only after the users ahead in the pipeline finish collecting water. There are nights when water pressure dips so low that some women get it after midnight. They split their day between household chores and collecting water.

Apart from the water scarcity caused by Coca-cola in Plachimada, the other districts in the state are facing a water crisis. For instance, in Kottayam district at some places, the water scarcity is so acute that people hesist to offer a glass of water to the visitor, which hitherto was a common custom. In the upper Kuttanadu area of the district during summer people collect water from a distance of 3-4 kms. Water supply from public taps is erratic and very often even after standing for an hour in the queue; people are not able to get a bucket of water. Most women and girls in Rajasthan find themselves searching water for much of the year. They trudge bare foot in the hot sun for hours over wastelands, across thorny fields, or rough terrain in search of water, often life the colour of mud and brackish, but still welcome for the parched throats back home. On an average, a rural woman walks more than 14000 km a year just to fetch water. Their urban sisters are only slightly better off- they do not walk such distances, but stand in the long winding queues for hours on end to collect water from the roadside taps on the water lorries.

In every household, in the rural areas in Rajasthan, women and girl children bear the responsibilit of collecting, transporting, storing, and managing water. In places, where there is no water for farming, men migrate to urban areas in search of work leaving women behind to fond for the old and the children. Women spend most of their time, collecting water with little time for other productive work. This impacts on the education of the girl child, if the girl is herself not collecting water, she is looking after the home and her siblings when her mother is away.

In brief:

- Water source being open dug well, the quality of water is poor; dirty, saline and has turbidity
- Women have to make at least three trips at 5 am, 11 am and 5 pm
- Sometimes, the number of trips is more
- Total distance traveled is 9-10 km, even higher

- Total Time spent is 6-9 hours
- Total number of pots/buckets is about 3 pots, 30-45 litres (one pot of 10-15 litres per trip)
- Due to long distance, they have to take rest in the middle of the way. Dust storms aggravate their problem
- At some villages water from tubewells is too saline to drink. Even animals particularly cow gets indigestion after drinking this water, so the villagers add water from the dug well. The entire life of women in rural areas like Jaisalmer is spent on water collection and cooking. Even the girls of 8-10 years cannot be spared. They cannot afford the luxury of school. For instance for Pappu a girl of hardly 10 years, water collection has become her main job. In the words of her grandmother "Water fetching is the schooling for Pappu." There are so many Pappus in the villages and dhaanis of Jaisalmer.

In Sriganganagar, the Indira Gandhi canal is the main source for drinking water. However, during the crisis period (either because of no water in the main canal/sub canal or due to the erratic power supply), the rich remain unaffected. In such crisis women from poorhouse hold draw water from the village diggis, which is totally unfit for any kind of human activity. They use this water not only for washing cloth and bathing but also for drinking. Due to the formation of algae, water becomes greenish and filthy. Women add alum to purify it.

In Orissa drinking water is being privatized. The government first insists on the formation of water associations and conveniently pass the responsibilities on to these association. When this proves inefficient, water distribution rights are given away to private contractors. For example, the Orissa government initially stressed on the formation of Paani Panchayats (water associations).

Later using police the government suppressed these panchayats justifying this by claiming that the villages were not being responsible enough. Titlagarh is the hottest town of India, but it has no water, causing great misery to the women. As the highest temperature, is recorded here 52 degrees centigrade which is also the highest temperature in India. People called "Titlagarh" as "Tatlagarh", in local language Tatla means hot. In Titlagarh water problem is acute. People are buying water throughout the year for drinking and cooking purpose. In the month of May and June the rate of water increase three times, from ₹ 2 per Dabba to ₹.8 per Dabba (container). This is the picture of urban areas, but in rural areas the problem is worse, where the tubewells all are becoming dry but people have no money to buy water. Due to the water problem some villagers are migrating to other places.

In Uttranchal women are suffering a lot in every village where water problem is severe. Natural sources are drying up which adds the kilometres for women everyday to quench the thirst of their family as well as animals.

Women are the major part of the workforce in Garhwal. They work from early morning to late evening to serve the family. They do all household work from cooking to cleaning and washing clothes and soiled utensils as well as look after their children and animals. Women also collect the water required for cooking, cleaning, washing, bathing and drinking both for human beings and animals.

During the survey in Jaunsar area of district Tehri Garhwal, in villages such as Nagthat, Duena, Vishoi, Gadol, Jandoh, Chi tar, Chichrad and Gangoa, it was observed that water in the region is mostly acidic in nature. The water problem in Chi tar and Gangoa villages is very severe, where men and women carry water on mules from 8-10 Km to the village. Because of the poor water quality, most of the villagers in the regions are suffering from many diseases related to skin and teeth. Natural resources of water in the area are very few and they are also disappearing very fast. During the survey, Smt. Nisha Devi of Chi tar village explained that they are not getting enough water for their animals so they take their animals to the spring about 2-3 kms away.

In Bundelkhand, women have no work but to collect drinking water on their heads from long distance. The grim situation of water may be best illustrated by one Bundelkhandi saying which roughly translated as “let the husband die but the earthen pot of water should not be broken”.

The scenario is worst in Patha in Chitrakut district where women have to travel a long distance to collect water for drinking. Half of the time of women is spent to collect water, which affects their health and the well-being of their children. The paucity of time due to water crisis aggravates the domestic problem.

Instead of solving the water crisis, attempts are being made to create a disastrous situation in the region. Banda city entirely depends on the Ken river. If Ken is linked with Betwa, then it will not affect only Banda, but would also jeopardize the survival of farmers who depend on Ken. Even in Delhi the water scenario is no better, being worst in Delhi slums. For example Sanjay colony slum of New Delhi has population about 15,000-20,000 with about 4500 households in the locality.

Majority of the population is self-employed and are engaged in making daris, mats and other clothes for sale. In this area people collect water from different sources depending on the availability such as DJB tanker, MCD pipe water supply and from Sulabh International. DJB tanker comes daily but it has no fixed time for water distribution. The water that comes from MCD pipe water has fixed time for water supply but it only comes for 1-2 hr in the evening (around 4.30 p.m.). At MCD pipe line people made bore and fetch water from it. If the people don’t get water from the above sources they are forced to get it from Sulabh International near Kalkaji temple for which they have pay @ ₹ 2 for 20 litre or so.

The Water crisis is same in West Bengal. In all the districts, the water commons have ceased to exist, and have become open-access resources, with hardly anyone responsible to take care of the resources. In North Bengal, some women reported they had opinions regarding the use of the water body, but their importance in management decisions is cipher. The absence of the community from the management of the water resources is indeed a tragedy, because now the resources are at the mercy of either the market or government officials.

Punjab; the name stands for abundance of water, but the present situation of water resources in the state is highly critical. The ground water availability is drastically hampered.

The village ponds are drying day by day. Women in the villages desperately need water. Near Talwandi Sabo, for some villages, the source for drinking water is about 8 km away. Near Jajjal due to contaminated water, women are suffering from a number of diseases including cancer. There have been several deaths attributed to polluted water.

For Maharashtra, water is an abiding concern. In many villages women have to walk more than 3 kilometres everyday to fetch two huge vessels of water illegally from a government reservoir. They have to make at least three trips everyday.

The state government do not send tankers to the villagers. At some places, women spend ₹ 5 for two canes of water. Images of women carrying the pots of water, walking miles and miles for one single pot are common in the state of Maharashtra. Women in Maharashtra have carried the water burden both as a result of scarcity and abundance.

Drought displacement due to dams and irrigation have contributed to increasing water burden of women. Women in Nandurbar district of North Maharashtra share their woes "forget about getting safe drinking water from wells, we spend most of our time locating streams and springs that quench our thrust". Many Women came as brides, their hair have gone dry, but the search for water has not ended. Karnataka is facing the worst kind of water crisis. In Bangalore, only 35 per cent of the city gets water on daily basis, the rest on alternative days.

In addition to the scarcity, erratic water supply is another problem. In Samadhanagar area, water generally comes in the morning at 11 A.M or in the middle of the night. Both these timings make it very difficult for women to collect water as they leave early in the morning to go to work. In Doddanagar slums in the city, women and children who are also breadwinners of the family spend 3-4 hours filling water, losing their wages. In Hosapalya locality women get severe joint pain in their shoulders, hips and knees due to carrying water pits from water sources outside their colony. In Peenya industrial area, many street fights occur among the women over water. Social conflict and tension is high due to water crisis. In brief, at an estimate about 150 Million-Woman Days and ₹ 10 Billion are lost in fetching water.

## CHALLENGES OF GLOBAL WATER

Productivity is a ratio between a unit of output and a unit of input. Here, the term water productivity is used exclusively to denote the amount or value of product over volume or value of water depleted or diverted.

The value of the product might be expressed in different terms (biomass, grain, money). For example, the so-called 'crop per drop' approach focuses on the amount of product per unit of water. Another approach considers differences in the nutritional values of different crops, or that the same quantity of one crop feeds more people than the same quantity of another crop. When speaking of food security, it is important to account for such criteria.

Another concern is how to express the social benefit of agricultural water productivity. All the options that have been suggested can be summarised by the phrases 'nutrient per drop', 'capita per drop', 'jobs per drop', and 'sustainable livelihoods per drop'. There is no unique definition of productivity and the value considered for the numerator might depend on the focus as well as the availability of data. However, water productivity defined as kilogram per drop is a useful concept when comparing the productivity of water in different parts of the same system or river basin and also when comparing the productivity of water in agriculture with other possible uses of water.

Crop water production is governed only by transpiration. As it is difficult to separate transpiration from evaporation from the soil surface between the plants (which does not contribute directly to crop production), defining crop water productivity using evapotranspiration rather than transpiration makes practical sense at field and system level. In irrigated agriculture in saline areas, the leaching requirement, *i.e.* the amount of water that needs to percolate to maintain rootzone salinity at a satisfactory level, should also be included together with evapotranspiration in the amount of water that is necessarily depleted during plant growth.

Other non-productive but beneficial uses could be included. Examples are evapotranspiration by windbreaks, cover crops, and the water used in wetting seedbeds to enhance germination.

The question of considering water losses from seepage and field percolation as consumption does not receive a unique response. If this water is of no use downstream or if it generates further pollution such as that resulting from geological salt leaching (*e.g.* San Joaquin Valley, California, the United States of America), then it must be accounted for as consumption.

Solutions to minimise these losses, such as canal lining or water improvement application, then have a positive effect on productivity. However, from a broader environmental point of view, it can be important to consider the impact of the outflow of an irrigation system on the overall productivity of an ecosystem.

As with the numerator, the choice of the denominator (which drops to be included) should depend on the scale, the point of view and the focus. At basin level, the choice might be between water diverted from the source and the same minus water restored, whereas at field level one might consider useful rain, irrigation water and supplemental irrigation.

## WATER PRODUCTIVITY IN ECONOMIC TERMS

Data are available for agricultural water productivity in economic terms for Jordan. Water productivity ranged from US\$0.3/$m^3$ for potato to US\$0.03/$m^3$ for wheat. The average value for agricultural products was US\$0.19/$m^3$ and for industrial products US\$7.5/$m^3$.

The IWMI analysed economic water productivity data from two irrigation systems in South Asia. The values for wheat production ranged from US\$0.07 to 0.17/$m^3$. Average systemwide water productivity values of US\$0.10 and 0.15/$m^3$ were reported for two other systems in South Asia. Systemwide values for a total of 23 irrigation systems in 11 countries in Asia, Africa and Latin America ranged from US\$0.03/$m^3$ (for a system in India) to US\$0.91/$m^3$ (for one in Burkina Faso), with an overall average of US\$0.25/$m^3$.

Comparison with the most recent cost of about US\$0.50/$m^3$ for desalinated seawater illustrates that this source of water is too expensive for virtually all agricultural production. However, its cost has come down to about one-tenth of what it was 20 years ago. Further improvements in the technology of seawater desalination are likely. Its cost is also likely to continue falling provided that as energy remains cheap.

## SPATIAL VARIABILITY OF WATER PRODUCTIVITY

Reported data on water productivity with respect to evapotranspiration ($WP_{ET}$) show considerable variation, *e.g.* wheat 0.6-1.9 kg/$m^3$, maize 1.2-2.3 kg/$m^3$, rice 0.5-1.1 kg/$m^3$, forage sorghum 7-8 kg/$m^3$ and potato tubers 6.2-11.6 kg/$m^3$, with incidental outliers obtained under experimental conditions. Data on field-level water productivity per unit of water applied ($WP_{irrig}$), as reported in the literature, are lower than $WP_{ET}$ and vary over an even wider range.

For example, grain $WP_{irrig}$ for rice varied from 0.05 to 0.6 kg/$m^3$, for sorghum from 0.05 to 0.3 kg/$m^3$ and for maize from 0.2 to 0.8 kg/$m^3$. The variability occurs because data were collected in different environments and under different crop management conditions. These affected the yield and the amount of water supplied.

Furthermore, it is often difficult to determine the real crop yield over a large area, *e.g.* the size of a large irrigation system. When asked for yield figures, individual farmers are likely to give a figure that depends on the situation. For a loan application, they may overstate the yield, whereas for payment of a debt or a tariff, they will probably understate the yield obtained. Vegetable yields of vegetables may change every day, and unless good records are kept, no one will know exactly how much was harvested during the total harvest period. Yields expressed in monetary terms are more doubtful as prices on the local market may fluctuate considerably over time.

Nevertheless, water productivity data across scales are useful in assessing whether water drained from upstream is reused effectively downstream. However, there are few reliable data on water productivity at different scale levels within the same system.

A study using remote sensing and GIS technologies assessed crop $WP_{ET}$ at various irrigation system scales in the Indus Basin in Pakistan. Crop water productivity was found to vary significantly at the scale of small canal command areas.

When water productivity was aggregated for canal command areas, the highest water productivity values decreased gradually. Their variability also decreased until at a scale of about 6 million ha water productivity tended to a low value of about 0.6 kg/m$^3$. This arose because at the larger scale, canal commands with less fertile or saline soils and with less canal water and poorer quality groundwater were included in the average.

## THE SUBSTANTIAL INCREASE OF WATER PRODUCTIVITY IN AGRICULTURE

Despite concerns about the technical inefficiency of water use in agriculture, water productivity increased by at least 100 per cent between 1961 and 2001. The major factor behind this growth has been yield increase. For many crops, the yield increase has occurred without increased water consumption, and sometimes with even less water given the increase in the harvesting index.

Example of crops for which water consumption experienced little if any variation during these years are rice (mostly irrigated) and wheat (mostly rainfed), for which the recorded increases worldwide amount to 100 and 160 per cent respectively. At the global level, the increase in water consumption for agriculture in the past 40 years has been 800 km$^3$ while world population has doubled to 6 000 million.

Considering that the arable rainfed area has not increased, one can conclude that with an additional 800 km$^3$ of water the world has been able to feed an additional 3 000 million people. This gives a rough estimate of 0.720 m$^3$/d/capita. This figure is low compared to the estimated global average for 2000 of 2.4 m$^3$/d/capita, which includes water for food at field level not including water losses. This is a good indicator of the significant productivity gain recorded in agriculture; a gain that has enabled the world to accommodate the doubling of the population and also increase intake.

As a whole, one can estimate that the water needs for food per capita halved between 1961 and 2001 from about 6 $m^3/d$ to less than 3 $m^3/d$.

The importance of water needs for food makes any small relative gain in this sector equivalent to a significant gain for other uses. For example, given the water needs for capita in 2000, a 1-per cent increase in water productivity in food production generates a potential of water use of 24 litres/d/capita. In order to produce the equivalent of the domestic water supply, a gain of 10 per cent in agricultural water productivity would be required, which is a matter of years.

Therefore, it can be argued that investing in agriculture and in agricultural water is the best avenue for freeing water for other purposes.

However, future agricultural gains will need to be split into several components: (i) compensation for the reduction of agricultural production areas as a result of urban encroachment, soil degradation, and the depletion of water resource availability or access (groundwater); (ii) increased water access for the rural poor and vulnerable groups; (iii) generation of wealthier farming systems; and (iv) freezing water for other uses including the environment.

## ENHANCING WATER PRODUCTIVITY AT PLANT LEVEL

Plant-level options rely mainly on germplasm improvements, *e.g.* improving seedling vigour, increasing rooting depth, increasing the harvest index (the marketable part of the plant as part of its total biomass), and enhancing photosynthetic efficiency. The most significant improvements in yield stability have usually resulted from breeding programmes to develop an appropriate growing cycle such that the duration of the vegetative and reproductive periods are well matched with the expected water supply or with the absence of crop hazards.

Planting, flowering and maturation dates are important in matching the period of maximum crop growth with the time when the saturation vapour pressure deficit is low. The periods of maximum crop growth may be optimised by means of breeding technology. Improved varieties with a deeper rooting system contribute to drought avoidance and the effective use of water stored in the soil profile.

Drought escape and increasing drought tolerance are also important strategies for increasing water productivity. Daylength-insensitive varieties of short to medium duration (90-120 d) enabled crops, such as wheat, rice and maize varieties developed as part of the green revolution, to increase water productivity by escaping late-season drought that adversely affects flowering and grain development. The modern rice varieties have about a threefold increase in water productivity compared with traditional varieties. Progress in extending these achievements to other crops has been considerable and will probably accelerate following the recent identification of the underlying genes. Genetic engineering, if properly integrated in breeding programmes and applied in a safe manner, can further contribute to the development of drought tolerant varieties and to increasing the water use efficiency.

## KEY PRINCIPLES FOR IMPROVING WATER PRODUCTIVITY

The key principles for improving water productivity at field, farm and basin level, which apply regardless of whether the crop is grown under rainfed or irrigated conditions, are: (i) increase the marketable yield of the crop for each unit of water transpired by it; (ii) reduce all outflows (*e.g.* drainage, seepage and percolation), including evaporative outflows other than the crop stomatal transpiration; and (iii) increase the effective use of rainfall, stored water, and water of marginal quality.

The first principle relates to the need to increase crop yields or values. The second one aims to decrease all 'losses' except crop transpiration. Its phrasing does not imply that it will be impossible to increase water productivity by reducing stomatal transpiration. It is conceivable that plant breeding may find ways to overcome this constraint. The third principle aims at making use of alternative water resources. The second and third principles should be considered parts of basinwide integrated water resource management (IWRM) for water productivity improvement. IWRM recognises the essential role of institutions and policies in ensuring that upstream interventions are not made at the expense of downstream water users.

## REAL IMPACTS OF VIRTUAL WATER ON WATER SAVINGS

Exchanges of virtual water through food trade first captured the attention of experts in the Near East, where water is scarce and imports represent considerable water savings. The value of virtual water of a food product is the inverse of water productivity. It is defined as the amount of water per unit of food that is or would be consumed during its production process.

Virtual water trade generates water savings for importing countries. It also generates global real water savings because of the differential in water productivity between the producing and the exporting countries.

For example, transporting 1 kg of maize from France (taken as representative of maize exporting countries for water productivity) to Egypt transforms an amount of water of about 0.6 $m^3$ into 1.12 $m^3$, which represents globally a real water saving of 0.52 $m^3$ per kilogram traded. In 2000, the maize imports in Egypt and the related virtual water transfer thus generated a global water saving of about 2 700 million $m^3$. The global real water saving is significant: a first estimate shows that water savings from virtual water transfer through food trade amounts to 385 000 million $m^3$.

Food storage also generates real water savings. For example, in the Syrian Arab Republic, 1988 was a good year for the cereal production with yields of 1.6 tonnes/ha, leading to a surplus. Thus, 1.9 million tonnes of cereals were stored during that year.

The following year was very dry, and the cereal yield dropped to 0.4 tonnes/ha. About 1.2 million tonnes of cereals were then withdrawn from storage to complement internal production and imports. Based on the water productivities recorded for these years, the estimated value of virtual water was to 1 and 3.33 $m^3$/kg respectively.

Therefore, the use of 1.2 million tonnes of cereals from storage in 1989 is equivalent to 4 000 million $m^3$ of virtual water. For the two-year period of reference (1988-89), some 2 800 million $m^3$ of water was saved by the food storage capacity. Globally, the trade in virtual water is rising rapidly. It increased in absolute value from about 450 $km^3$ in 1961 to 1 340 $km^3$ in 2000, reaching 26 per cent of the total water required for food including equivalence for sea products and sea fish. This value is shared evenly between energy, fat and protein products.

## RAISING WATER PRODUCTIVITY AT FIELD LEVEL

Improved practices at field level relate to changes in crop, soil and water management. They include: selecting appropriate crops and cultivars; planting methods (*e.g.* on raised beds); minimum tillage; timely irrigation to synchronise water application with the most sensitive growing periods; nutrient management; drip irrigation; and improved drainage for water table control.

Water depletion occurs when water evaporates from moist soil, from puddles between rows and before crop establishment. All cultural and agronomic practices that reduce these losses, such as different row spacings and the application of mulches, improve water productivity. The irrigation method also affects these evaporative losses. Drip irrigation causes much less soil wetting than sprinkler irrigation. The significance of soil improvement in enhancing water productivity is often ignored. However, integrated crop and resource management practices, such as improved nutrient management, can increase water productivity by raising the yield proportionally more than it increases evapotranspiration. This principle applies to both irrigated and rainfed agriculture. Integrated weed and integrated pest management have also contributed effectively to yield increases.

One of the field-level methods for increasing water productivity is deficit irrigation, where deliberately less water is applied than that required to meet the full crop water demand. The prescribed water deficit should result in a small yield reduction that is less than the concomitant reduction in transpiration.

Therefore, it causes a gain in water productivity per unit of water transpired. In addition, it could lower production costs if one or more irrigations could be eliminated. For deficit irrigation to be successful, farmers need to know the deficit that can be allowed at each of the growth stages and the level of water stress that already exists in the rootzone. Most importantly, they need to have control over the timing and amount of irrigations. Deficit irrigation carries considerable risk for the farmers where water supplies are uncertain, as is the case with rainfall or unreliable irrigation supplies. Where water availability falls below a certain level, the value of the crop can fall to zero, either because the crop dies or because the product is of such low quality as to be unmarketable. When water is scarce, farmers could reduce the irrigation as appropriate to maximise returns to water if they have control over the timing and amount of irrigations.

This degree of flexibility is usually the case with sprinkler and drip irrigation, and also with pumped groundwater if the farmer owns the pump. A totally flexible delivery system for surface irrigation in large irrigation systems is expensive because of the required overcapacity in the conveyance system.

The trade-off between reduced yield and higher water productivity needs to be quantified in economic terms before recommending deficit irrigation (and other water-saving irrigations in rice production).

The often cited low water productivity per unit of water supply in rice cultivation derives from considering as losses the percolation resulting from the standing water layer on the field surface. However, this water is often recycled, and rice water productivity generally compares well with that of a dry cereal. Nevertheless, water-saving irrigation techniques such as saturated soil culture and alternate wetting and drying can reduce the unproductive water outflows drastically and increase water productivity. These techniques generally lead to some yield decline in the current lowland rice high-yielding varieties. However, some experiments are reporting substantial yield increases for local varieties using a technique called system rice intensification (SRI), a technique which originated in Madagascar. Here again there is no unique response; the fit with local resources and capacity is the most important feature to account for.

Without anticipating results of current investigations in many countries, it seems that the potential of the SRI technique for the poor to increase the productivity of scarce land and water is significant provided that enough family labour is available. Other approaches are being researched as part of efforts to increase water productivity without sacrificing yield. One of these is to develop so-called aerobic rice systems that allow rice cultivation in non-flooded conditions. The development of these new rice varieties is essential if rice is to be grown like other irrigated upland crops and the deep percolation associated with paddy rice is to be avoided.

# 4

# Soil Conservation

## INTRODUCTION

Soil conservation is a set of management strategies for prevention of soil being eroded from the Earth's surface or becoming chemically altered by overuse, acidification, salinization or other chemical soil contamination.

It is a component of environmental soil science. Decisions regarding appropriate crop rotation, cover crops, and planted windbreaks are central to the ability of surface soils to retain their integrity, both with respect to erosive forces and chemical change from nutrient depletion.

Crop rotation is simply the conventional alternation of crops on a given field, so that nutrient depletion is avoided from repetitive chemical uptake/ deposition of single crop growth.

## EROSION PREVENTION

There are also conventional practices that farmers have invoked for centuries. These fall into two main categories: contour farming and terracing, standard methods recommended by the U.S. Natural Resources Conservation Service, whose Code 330 is the common standard.

Contour farming was practiced by the ancient Phoenicians, and is known to be effective for slopes between two and ten per cent. Contour plowing can increase crop yields from 10 to 50 per cent, partially as a result from greater soil retention.

There are many erosion control methods that can be used such as conservation tillage systems and crop rotation. Keyline design is an enhancement of contour farming, where the total watershed properties are taken into account in forming the contour lines. Terracing is the practice of creating benches or nearly level layers on a hillside setting. Terraced farming is more common on small farms and in underdeveloped countries, since mechanized equipment is difficult to deploy in this setting. Human overpopulation is leading to destruction of tropical forests due to widening practices of slash-and-burn and other methods of subsistence farming necessitated by famines in lesser developed countries. A sequel to the deforestation is typically large scale erosion, loss of soil nutrients and sometimes total desertification.

## PERIMETER RUN-OFF CONTROL

Trees, shrubs and groundcovers are also effective perimeter treatment for soil erosion prevention, by insuring any surface flows are impeded. A special form of this perimeter or inter-row treatment is the use of a "grassway" that both channels and dissipates run-off through surface friction, impeding surface run-off, and encouraging infiltration of the slowed surface water.

# SOIL EROSION AND ITS CONSERVATION

Soil is one of the most important natural resources of man. Soils are essential for man for growing crops, fodder and limber. Once the fertile portion of the earth's surface is lost, it is very difficult to replace it. In India, the destruction of the top-soil has already reached an alarming proportion. Land degradation problems have resulted in increasing depletion of the productivity of the basic land stock through nutrient deficiencies.

In addition to the direct loss of crop producing capacity, soil erosion increases the destructiveness of floods and decreases the storage capacity of water in reservoirs. It is therefore essential that the soils should not be allowed to wash or blow-away more rapidly than they can be regenerated, their fertility should not be exhausted and their physical structure should remain suited to continued production of desired plant materials.

Protection of land from further degradation, adoption of various conservation measures, including reclamation and scientific manage-ment of available land stock is very important for a country like India to achieve higher productivity of food, fodder, fuel and industrial raw materials on a substantial basis. Besides, demand for land for providing social priorities such as shelter, roads, industrial activities is increasing at a very fast rate with the rise in population and very often good agricultural and forest lands are being diverted to such use. It is, therefore, necessary to keep soil in place and in a state favourable to its highest productive capacity.

## SOIL EROSION

The process of destruction of soil and the removal of the destroyed soil material constitute soil erosion. According to Dr. Bennett "the vastly accelerated process of soil removal brought about by the human interference, with the normal disequilibrium between soil building and soil removal is designated as soil erosion".

## TYPES OF SOIL-EROSION

Erosion of soil by water is quite significant and takes place chiefly in two ways (a) Sheet erosion, (b) Gully erosion.

## SHEET EROSION

Sheet movement of water causes sheet erosion and depends on the velocity and quantity of pronounced surface run-off and the erodability of the soil itself. In such cases, the soil is eroded as layers from the hill slopes, sometimes slowly and in-sidiously and sometimes more rapidly.

*Sheet erosion is more or less universal on*:

- All bare follow land,
- All uncultivated land whose plant cover has been thinned out by over grazing, fire or other misuse, and
- All sloping cultivated fields and on sloping forest, scrub jungles where natural porosity of soil has been removed by heavy grazing, felling of trees or burning etc.

The particles loosened and shifted by the rain drops are carried down slope by a very thin sheet of water which moves along the surface. The impacts of the raindrops increases the turbulance and transporting capacity of this unchannelized sheetwash which results in the uniform skimming of the top soil. Sheet erosion is considered as dangerous as it may continue for years but may or may not leave any trace of the damage. Sheet erosion is common in the Himalayan foot-hills, in Assam, Western ghats and Eastern ghats. When sheet erosion continues unchecked, the silt laden run-off forms well-defined minute finger shaped grooves over the entire field. Such thin channeling is known as 'rill-erosion', which is active over wide areas in Bihar, Uttar Pradesh, Madhva Pradesh and in semiarid areas of Maharashtra, Karnataka, Andhra Pradesh and Tamil Nadu.

## GULLY EROSION

On a gentle slope, adequately covered by vege-tation, clay soil will resist erosion to a great extent and the water forms small rivulets which can then erode deeper. The rivulets in turn join together to form larger channels until gullies are formed gradually deep gullies cut into the soil and then spread and grow until all the soil is removal from the sloping ground.

This phenomenon once started and if not checked, goes on extending and ultimately the whole land is converted into a bad-land topography. Gully erosion is more common in areas where the river system has cut down into elevated plateaus so that feeders and branches carve out an intricate pattern of gullies. Apart from this, it also takes place in relatively level country whenever large blocks of cultivation give rise to con-centration of field run-off.

## WIND EROSION

It occurs in dry climatic areas having a sparse and low vegetation cover on mechanically weathered, loosened surficial material. Dust storms are the principal agents of wind erosion. The top soil is often blown off from the surface rendering it infertile. Besides, with the decrease in the wind velocity coarse sand particles get deposited in some areas covering the existing soil and rendering it unproductive.

## SOIL EROSION AND CONSERVATION

One of the most crucial aspects of soil erosion research is the very conceptualization of why soil erosion is a socioeconomic problem. Traditionally this has not been an issue, since sustained high levels of soil loss were presumed to irreversibly degrade the productivity of agricultural resources. Accordingly, if the soil erosion problem is conceptualized as medium to long-term land productivity decline detrimental to the interests of farmers, it would follow that conservation should be achieved through voluntary farmer compliance (rather than through mandatory regulation) based on the logic of the long-term interests of farmers.

Historically, federal and state government policy thus has emphasized education of farmers and modest levels of financial inducements (such as Agricultural Stabilization and Conservation Service cost-sharing programmes) to lever farmer decision making. But it has been increasingly recognized that the proportion of prime U.S. farmland that stands to be irrevocably degraded by soil erosion is relatively small and that the "off-site" costs of soil erosion — the impacts of soil erosion and run-off on water and land resources of a farmer's parcel — are very significant and may very well be in excess of the on-site costs.

These alternative conceptualizations of the soil erosion problem — productivity loss versus destruction of off-site water and land resources — have very profound implications for appropriate types of social science research and for public policy. To the degree that soil erosion primarily results in land productivity losses, research should focus on individual- and farm-level factors that influence resource management, and public policy should incorporate these research findings in order to develop educational and incentive programmes to achieve soil conservation through farmer self-interest.

To the degree that soil erosion is primarily a problem because of off-site costs, farmers cannot be expected to conserve because the long-term productivity benefits of conservation are small, and research at a more macro level (*e.g.*, to determine the degree to which farmers' incentive structures and resource management behaviours are congruent with the public's interests in clean, navigable waters) would be most appropriate.

Debate over the appropriate kind of analysis in soil conservation research has been closely related to, and in many respects was stimulated by, debate over the adoption-diffusion approach. Pampel and van Es, reported that the correlates of adoption of conservation technologies tended to be different from those of commercial non-conservation technologies. They also suggested that effective conservation practices will tend not to be profitable for farmers, so that to achieve the public interest in soil conservation voluntary compliance among farmers may be insufficient.

The issues raised by Pampel and van Es have continued to pervade the erosion and conservation literature. Many researchers have conducted their research by seeking to revise the microsociological diffusion of innovations approach, while others have argued that the constraints to soil conservation behaviour tend to be more macro in nature and that the diffusion of innovations approach will be inadequate.

Prior to 1970 soil and water conservation technologies and practices were generally understood to be a diverse collection of cropping patterns (crop rotations, contour planting, strip cropping, cover cropping, sod waterways, filter strips) and physical and biomass structures (terraces, sediment retention basins, diversion channels, animal waste structures, and hedge rows). Since that time, however, soil and water conservation has, for the majority of researchers and policymakers, come to be largely conterminous with the adoption and use of no-till and related conservation and reduced tillage equipment, which enables farmers to leave a mulch layer of residue on the soil to reduce run-off and sediment losses.

Two observations can be made with regard to traditional and reduced-tillage equipment approaches to soil and water conservation in agriculture. First, these two approaches, while by no means incompatible, are nonetheless quite different and will tend to apply most appropriately to different types of farm operations. Traditional cropping practices and structures are often most attractive to smaller, more diversified farmers, while reduced tillage tends to be most attractive to larger, more highly mechanized producers of row crops.

Second, reduced tillage, although acknowledged to be effective in reducing erosion in row-crop monocultures, has become somewhat controversial. One controversial aspect of the reduced-tillage approach has been that it requires increased use of herbicides, which may lead to contamination of water and thereby negate some of the run-off reduction benefits of reduced tillage. Many studies have indicated, moreover, that reduced tillage practices have been adopted by farmers largely because of labour savings rather than because of the ability of the technology to reduce soil erosion.

Indeed, the farmers who have been most likely to adopt no-till and other reduced tillage technologies have been found to be large operators with highly specialized operations involving practices such as continuous cropping and row-crop monocultures.

Thus farmer-adopters of such conservation technologies ironically represent the major trends in U.S. farm structure that have been demonstrated to lead to environmental degradation in agriculture.

However, farmer perceptions of a threat to soil productivity may lead to adoption of soil-conserving innovations, whereas stewardship related motivations may have little effect on motivations to prevent other kinds of consequences such as off-site pollution.

## SOIL NATURAL RESOURCES AND CONSERVATION SERVICE

The Natural Resources Conservation Service (NRCS), formerly known as the Soil Conservation Service (SCS), is an agency of the United States Department of Agriculture (USDA) that provides technical assistance to farmers and other private landowners and managers.

Its name was changed in 1994 during the Presidency of Bill Clinton to reflect its broader mission. It is a relatively small agency, currently comprising about 12,000 employees. Its mission is to improve, protect, and conserve natural resources on private lands through a cooperative partnership with local and state agencies.

While its primary focus has beenagricultural lands, it has made many technical contributions to soil surveying, classification and water qualityimprovement. One example is the Conservation Effects Assessment Project (CEAP), set up to quantify the benefits of agricultural conservation efforts promoted and supported by programmes in the Farm Security and Rural Investment Act of 2002 (2002 Farm Bill). NRCS is the leading agency in this project.

### PROGRAMMES AND SERVICES

As of November 2011 the NRCS consists of 42 programmes and activities. The NRCS also offers services to private land owners, conservation districts, tribes, and other types of organizations.

### FARM BILL

The conservation provisions in the Food, Conservation, and Energy act of 2008. This bill provides conservation opportunities for farmers. Services also include financial assistance which makes sure money is allocated and used properly in providing conservation for various natural resources, technical assistance which is given through the Conservation Technical Assistance programme (CTA). This service is available to anyone interested in conservation of natural resources and easements. This includes:

### FARM AND RANCH LAND PROTECTION PROGRAMME

(FRPP) The purpose of this programme is to work with land owners to purchase development rights to current farm and ranch land, in order to keep said land from being developed for other uses. The programme matches funds from property owners, and can be applied whether potential buyers are private or from state or local government, as well as Native American tribes. In order to receive funds from the programme, the land must be privately owned, and have an offer for sale pending. The land must be large enough to support substantial agricultural yield, and be surrounded by land with a similar nature. If there is a danger for soil erosion, a conservation plan must be included.

## GRASSLANDS RESERVE PROGRAMME

(GRP) Volunteer programme to increase animal and plant biodiversity, and to protect grasslands. Participants limit use of grassland for commercial and agricultural development. The land may still be grazed or seeded, with the exception of the nesting seasons of bird species that are protected under law. A grazing management plan must be submitted for participation.

## HEALTHY FORESTS RESERVE PROGRAMME

(HFRP) Landowners volunteer to restore and protect forests in 30 or 10 year contracts. This programme hands assisting funds to participants.

*The objectives of HFRP are to:*

- Promote the recovery of endangered and threatened species under the Endangered Species Act (ESA)
- Improve plant and animal biodiversity
- Enhance carbon sequestration.

## WETLANDS RESERVE PROGRAMME

(WRP) Volunteer programme for landowners to protect or restore wetlands on properties they own. The programme offers both financial and technological support to these landowners in order to help cultivate long term wetland health with optimal biodiversity per acre.

## NRCS NATIONAL AG WATER MANAGEMENT TEAM

(AGWAM) Serves 10 states in the Midwest United States in helping to reduce Nitrate levels in soil due to run-off from fertilized farmland. The project began in 2010 and initially focused on the Mississippi Basin area. The main goal of the project is to implement better methods of managing water drainage from agricultural uses, in place of letting the water drain naturally as it had done in the past. In October 2011, the The National "Managing Water, Harvesting Results" Summit was held to promote the drainage techniques used in hopes of people adopting them nationwide.

## SNOW SURVEY AND WATER SUPPLY FORECASTING

Includes water supply forecasts, reservoirs, and the Surface Water Supply Index (SWSI) for Alaska and other Western states. NRCS agents collect data from snowpack and mountain sites to predict spring run-off and summer streamflow amounts. These predictions are used in decision making for agriculture, wildlife management, construction and development, and several other areas. These predictions are available within the first 5 days of each month from January to June.

## WILDLIFE HABITAT INCENTIVE PROGRAMME

(WHIP) Is a volunteer programme for improving habitats for wildlife on farmlands, private lands, and Indian land. The programme was renewed in 2008 under the Food, Conservation, and Energy Act. WHIP provides up to 75 per cent cost share and technical assistance, and includes both terrestrial and freshwater aquatic habitats.

*The programme aims to protect seven threatened species in particular*:

- Lesser Prairie Chicken
- New England Cottontail
- Southwestern Willow Flycatcher
- Greater Sage-Grouse
- Gopher Tortoise
- Bog Turtle
- Golden-winged Warbler

## CONSERVATION TECHNICAL ASSISTANCE PROGRAMME

(CTA) Is a blanket programme which involves conservation efforts on soil and water conservation, as well as management of agricultural wastes, erosion, and general longterm susta-inability. NRCS and related agencies work with landowners, communities, or developers to protect the environment. Also serve to guide people to comply with acts such as the Highly Erodible Land, Wetland (Swampbuster), and Conservation Compliance Provisions acts. The CTA can also cover projects by state, local, and federal governments.

## GULF OF MEXICO INITIATIVE

(GoMI)Is a programme to assist gulf bordering states (Alabama, Florida, Louisiana, Mississippi, and Texas) improve water quality and use sustainable methods of farming, fishing, and other industry. The programme will deliver up to 50 million dollars over 2011-2013 to apply these sustainable methods, as well as wildlife habitat management systems that do not hinder agricultural productivity, and prevent future over use of water resources to protect native endangered species.

## INTERNATIONAL PROGRAMMES

The NRCS (formerly SCS)has been involved in soil and other conservation issues internationally since the 1930s. The main bulk of international programmes focused on preventing soil erosion by sharing techniques known to the United States with other areas. NRCS sends staff to countries worldwide to conferences to improve knowledge of soil conservation.

There is also international technical assistance programmes similar to programmes implemented in the United States. There are long term technical assistance programmes in effect with one or more NRCS staff residing in the country for a minimum of one year. There are currently long term assistance programmes on every continent. Short term technical assistance is also available on a two week basis.

These programmes are to encourage local landowners and organizations to participate in the conservation of natural resources on their land, and lastly landscape planning has a goal to solve problems dealing with natural resource conservation with the help of the community in order to reach a desired future outcome.

## TECHNICAL RESOURCES

## WATER

Pollution of water due to a number of different pollutants has driven the NRCS to take action. Not only do they offer financial assistance but they also provide the equipment needed for private land owners to protect our water resources. Water gets polluted by nitrogen and phosphorus which causes algae to grow proliferously causing the oxygen concentrations to decline rapidly, life is no longer supported in this habitat.

Excessive sedimentation is also another concern along with pathogens threats that can find their way into water systems and cause detrimental effects. NRCS works in a way to help both the land owner and the water systems that need prevention or restoration.

## WATER MANAGEMENT

Water management strives to manage and control the flow of water in a way that is efficient while causing the least amount of damage to life and property.

This helps provide protection in high risk areas from flooding. Irrigation water management is the most efficient way to use and recycle water resources for land owners and farmers. Drainage management is the manipulation of sub surface drainage networks in order to properly disperse the water to the correct geographical areas. The NRCS engineering division is constantly making improvements to irrigation systems in a way that incorporates every aspect of water restoration.

## WATER QUALITY

A team of highly trained experts on every aspect of water is employed by the NRCS to analyze water from different sources. They work in many areas such as: hydrology and hydraulics, stream restoration, wetlands, agriculture, agronomy, animal waste management, pest control, salinity, irrigation, and nutrients in water.

## WATERSHED PROGRAMME

Under watershed programmes the NRCS works with states, local governments, and tribes by providing funding and resources in order to help restore and also benefit from the programmes.

They would like to provide: watershed protection, flood mitigation, water quality improvement, soil erosion reduction, irrigation, sediment control, fish and wildlife enhancement, wetland and wetland function creation and restoration, groundwater recharge, easements, wetland and floodplain conservation easements, hydropower, watershed dam rehabilitation.

## PLANTS AND ANIMALS

Plants and animals play a huge role in the health of our ecosystems. A delicate balance exists between relationships of plants and animals. If an animal is introduced to an ecosystem that is not native to the region that it could destroy plants or animals that should not have to protect itself from this particular threat.

As well as if a plant ends up in a specific area where it should not be it could have adverse effects on the wildlife that try to eat it. NRCS protects the plants and animals because they provide us with food, materials for shelter, fuel to keep us warm, and air to breathe. Without functioning ecosystems we would have none of the things mentioned above. NRCS provides guidance to assist conservationists and landowners with enhancing plant and animal populations as well as helping them deal with invasive species.

## FISH AND WILDLIFE

NRCS for years has been working towards restoration, creation, enhancement, and maintenance for aquatic life on the nearly 70 per cent of land that is privately owned in order to keep the habitats and wildlife protected. NRCS with a science based approach, provides equipment to wildlife and fish management. They also do this for landowners who qualify to benefit from these technologies.

## INSECTS AND POLLINATORS

Pollination via insects plays a huge role in the production of food crop and flowering plants. Without pollinators searching for nectar and pollen for food the plants would not produce a seed that will create another plant. NRCS sees the importance of this process so they are taking measures to increase the declining number of pollinators.

There are many resources provided from the NRCS that will help any individual do their part in conservation of these important insects. Such as

Backyard Conservation which tells an individual exactly how to help by just creating a small habitat in minutes There are many others such as: Plants for pollinators, pollinators habitat in pastures, pollinator value of NRCS plant releases in conservation planting, plant materials publications relating to insects and pollinators, PLANTS database: NRCS pollinator documents. All of these are valuable resources that any individual can take advantage of.

## INVASIVE SPECIES AND PESTS

Many adverse effects are present due to invasive species. Plants and animals both inhabit areas that they are not intended to be. The kudzu vine for example covers miles of foliage. These invasive species cause America's reduction in economic productivity and ecological decline. Humans are unknowingly transporting these invasive species via ships, planes, boats, and their own bodies. NRCS works in collaboration with the plant materials centers scattered throughout the country in order to get a handle on the invasive species of plants.

These centers scout out the plants and take measures to control and eradicate them from the particular area.Invasive animals such as feral hog, european gypsy moth, and the sirex woodwasp pose a significant threat to America's wildlife as well as to the health of human beings.

The hog was introduced as a food source for humans, but now the swine pandemic is a serious threat to humans. They gypsy moth destroys natural forests that are habitat to many beneficial species. The Woodwasp feeds on pine trees as well as providing a means of transportation for a fungus that kills pine trees.

## LIVESTOCK

Livestock management is an area of interest for the NRCS because if not maintained valuable resources such as food, wools, and leather would not be available. The proper maintenance of livestock can also improve soil and water resources by providing a waste management system so that run off and erosion is not a problem.

The NRCS provides financial assistance to land owners with grazing land and range land that is used by livestock in order to control the run off of waste into fresh water systems and prevent soil erosion.

## PLANTS

Plants are a huge benefit to the health of ecosystems. NRCS offers significant amounts of resources to individuals interested in conserving plants. From databases full of information to financial assistance the NRCS works hard to provide the means needed to do so.

The plant materials programme, Plant materials centers, Plant materials specialists, PLANTS database, National Plant Data Team (NPDT) are all used together to keep our ecosystems as healthy as possible. This includes getting rid of unwanted species and building up species that have been killed off that are beneficial to the environment. The NRCS utilizes a very wide range of interdisciplinary resources.

*The NRCS also utilizes the following disciplines in order to maximize efficiency:*

- Agronomy
- Erosion
- Air Quality and Atmospheric Change
- Animal Feeding Operations and Confined Animal Feeding Operations
- Biology
- Conservation Innovation Grants
- Conservation Practices
- Cultural Resources
- Economics
- Energy
- Engineering
- Environmental Compliance
- Field Office Technical Guide
- Forestry,
- Maps
- Data and Analysis
- Nutrient Management
- Pest Management
- Range and Pasture
- Social Sciences
- Soils, and Water Resources

These Science-Based technologies are all used together in order to provide the best conservation of natural resources possible.

## TILLAGE SYSTEMS IN SOIL CONSERVATION

Conservation tillage systems cover a broad spectrum of farming methods primarily aimed at conserving and managing crop residue to reduce erosion. In the past, a residue cover of 30 per cent was considered adequate for erosion control and became the goal. Experience and data have shown, however, that truly effective erosion control relies on a situation- and site-specific plan developed by the individual.

Because 30 per cent was the former definition of conservation tillage, research data used it as a reference point or standard. Crop residue management is becoming the focus when looking at tillage systems for erosion control.

Now 30 per cent is just a point along the continuum and may or may not be the objective for a given situation. The tillage system of choice is the single largest factor in effective residue management. Each additional field operation incorporates more residue and increases erosion potential.

*Selecting the most appropriate system for a particular soil and cropping situation requires matching the operations to:*

- Crop sequence.
- Topography.
- Soil type.
- Drainage.
- Weather conditions.

Tillage systems such as ridge-till and no-till leave more residue and offer greater erosion control. No-till is the only system that consistently leaves at least 30 per cent cover following soybeans. For residue management, rotating tillage systems to coincide with crop rotations is an option.

For example, a notill system following soybeans and a chisel or disk system following corn can provide adequate erosion control after soybeans and allows for some tillage in the less fragile and more abundant corn residue. Rotating tillage systems, however, slows the development of improved soil structure and may require investment in more equipment.

To minimize equipment needs, the selection of a tillage system should consider all crops in the crop rotation across all the soils of the farm. However, there may be some soils or some crops which may perform better with a different tillage system than the system selected for the rest of the farm. In these situations, a careful machinery analysis needs to be performed to see if the yield benefits would cover the additional fixed and variable costs of two or more tillage and planting systems. Sometimes, this analysis justifies the additional equipment, such as with no-till drilled soybeans and/or wheat combined with a mulch-till corn. Other times, it may be desirable to use only one tillage system, such as with ridge-planting with continuous row crops.

## TILLAGE SYSTEM DESCRIPTIONS

### CHISEL PLOW

The chisel plow produces a rough surface and generally leaves 50 per cent-70 per cent of the existing corn or grain sorghum residue on the surface depending on chisel point selection, shank spacing, operating speed and depth. Straight, narrow points, about 2" wide, leave the most residue. However, following crops that leave fragile residue even narrow points bury too much residue to provide adequate protective cover.

Where erosion is not a primary concern, 3" or 4" wide, twisted points invert more soil and bury more residue. In the western Corn Belt and the High Plains, wide sweeps are sometimes used on a chisel plow in wheat residue to undercut weeds and leave residue on the surface.

Typically, chiseling is performed in the fall and is followed by one or more secondary tillage operations in the spring. The fall operation cuts and incorporates some of the residue, making it more susceptible to decomposition and overwinter weathering than undisturbed residue. Partially decomposed residue is easily broken and covered by secondary tillage operations, negating much of the effect of having selected chisel points which leave more residue.

On many soils, a single pass in the spring with a disk, field cultivator or combination tillage implement provides limited pesticide and fertilizer incorporation on fall chiseled fields. A second tillage pass provides more complete incorporation, but can decrease residue and erosion control. Spring chiseling affords erosion control during the winter and allows extended grazing of stalks. However, soil moisture evaporation following spring tillage can result in yield reductions, particularly in lower rainfall areas. Spring chiseling may also produce clods that could require additional tillage operations to prepare a suitable seedbed. Compaction problems may also arise on wet soils with multiple tillage trips. Larger implements may be required to provide timeliness with spring tillage operations, especially in wet springs.

A chisel plow may clog in extremely heavy or wet residue unless stalk shredding or light tillage precedes chiseling. This additional operation increases fuel and labour requirements. Several combination tillage implements have coulters or disks mounted in front of the chisel shanks which often eliminate the need for a prechiseling operation.

The coulters or disks are operated just deep enough to cut the surface residue. This reduces the chance for residue clogging in the chisel area. The coulters will redistribute residue having the effect of increasing the percentage of ground covered. The shank spacing on these machines is usually 15" compared to 12" for a conventional chisel plow.

## DISK AND FIELD CULTIVATION

The tandem disk harrow is the most commonly used tillage implement in the Corn Belt. Typically, the disk harrow is followed by a field cultivator for final seedbed preparation.

About 40 per cent-70 per cent of the residue generally remains on the surface after a single disking of corn, grain sorghum or wheat residue. Generally, disking corn or grain sorghum residue more than twice buries too much residue for effective erosion control. One disking is adequate if using a field cultivator for final seedbed preparation. However, in fragile residue, even a single pass with any commonly used tillage implement does not leave enough residue for effective erosion control.

Fall disking saves time in the spring but the erosion potential from wind and rain increases, and snow entrapment decreases. The tilled residue is exposed to weathering, becoming more fragile and more easily destroyed with subsequent tillage operations. A spring disk system minimizes erosion during the winter and is well suited to adequately drained and lighter textured soils.

A disk incorporates herbicides and other surface applied products. However, this incorporation occurs at the expense of soil structure. A common problem is disking when soils are too wet. Disking wet soils results in non-uniform incorporation, creates clods that require additional tillage operations and leaves a compacted soil layer below the depth of disking that can restrict root growth and reduce yields, especially in dry years. Disking wet soils to dry them out does dry the surface layer but creates considerable compaction below the depth of tillage.

A field cultivator accomplishes secondary tillage imme-diately preceding planting. The amount of residue covered by a field cultivator depends on the amount of time and weathering since the primary tillage operation. If field cultivation occurs two to three days after primary tillage, the field cultivator may not reduce the residue any further.

In some circumstances, particularly with corn residue, field cultivation soon after primary tillage redistributes residue and may slightly improve per cent coverage. On the other hand, if time between primary tillage and field cultivation allows for some decomposition, a field cultivator significantly reduces residue.

One pass of a field cultivator or combination tillage implement is an alternative to a disk. The most common onepass tillage system is used in soybean residue. While a one-pass system is successful in terms of yield and reduced costs, there is not enough soybean residue to effectively reduce soil losses on fields prone to erosion.

## STUBBLE MULCH

The blade plow, or sweep plow, a common tillage implement in the High Plains, cuts weeds at the roots and leaves most of the residue anchored at the surface with minimum disturbance of the soil surface. Blade plowing is typically a summer fallow operation after small grain harvest. It kills weeds and loosens the surface. In moist soils, particularly those with higher clay contents, a blade plow may cause soil smearing below the blade, thereby limiting its use as a spring tillage implement.

The blade plow should be used on a hot day when the soil is dry for maximum weed kill. Small weeds are lifted and uprooted so they dry out. Deep operation often leaves roots in the soil and weeds re-root. Attaching a rod weeder or mulch treader behind a blade plow to bring roots to the surface improves weed control. Using these attachments, especially the mulch treader, results in greater residue loss and soil pulverization.

Using either a blade plow or rod weeder at progressively shallower depths provides a firm seedbed while controlling subsequent weed growth. These implements work beneath the soil surface and are typically used in drier climates; 30 per cent or more residue can usually be maintained. However, substituting either a disk or field cultivator for secondary tillage greatly decreases the probability of leaving adequate residue for erosion control.

Using herbicides for weed control and planting after one pass of the blade plow reduces or eliminates secondary tillage operations, but causes problems at planting because the soil is loose and fluffy, making it difficult to cut the loosened residue.

## RIDGE-TILL

In ridge-till, crops are planted into ridges formed during cultivation of the previous crop. Ridge cleaning devices push residue and surface weed seeds off the ridge either during planting or during a separate, preplant operation. A band application of herbicide behind the planter provides weed control in the row. Crop cultivation controls weeds between the rows and rebuilds the ridges for the following year.

Ridge-till reduces erosion by leaving the soil covered with residue until planting. After planting, 30 per cent-50 per cent residue may be left, but it is not uniformly distributed. Residue-covered areas between the rows alternate with residue-free strips in the row area. For erosion control, the ridges should be 3"-5" higher than the furrows after planting and that ridges be shaped to shed water to the furrow. For the most effective erosion control, orient ridges approximately on the contour.

Crop rotation influences the suitability of ridge systems. Ridges are maintained year-to-year with a cultivator, making ridge-till well suited to continuous row crops. Two cultivations are generally required: the first loosens soil and controls weeds, the second provides additional weed control and rebuilds the ridges. For ease of planting, the ridges should be rounded or flat topped, and 6"-8" tall after cultivation. Proper ridge shape and annual maintenance are keys to a successful ridge system. Be careful not to damage or destroy the ridges by wheel traffic, particularly during harvest.

Level or gently sloping fields, especially those with poorly drained soils, are well suited to ridge systems. The elevated ridges warm earlier in the spring. This warming, combined with drainage from the ridge, allows soil in the ridge to be drier at planting than untilled, unridged soil. A ridge system is an excellent choice for soils that are often too wet for early spring tillage, especially in the northern Corn Belt where the growing season is shorter.

Ridge systems complement furrow irrigation. Ditching, furrowing or hilling for irrigation provides suitable ridges for planting the following year. Chopping stalks or performing a very shallow, high speed tillage operation removes residue from the ridges and aids furrow irrigation, especially on soils with higher infiltration rates. In a ridge-plant system, row cleaning devices on the planter move a small amount of soil, residue and weed seed off the ridge top. Ridge-cleaning attachments include sweeps, disk furrowers or horizontal disks. Except for possible fertilizer injection, no soil disturbance occurs prior to ridge planting.

In contrast, some ridge-till systems use limited tillage prior to planting. Tillage is generally very shallow, disturbing only the ridge tops. This tillage smooths peakshaped ridges to help keep the planter on the old rows. Depending on the tillage implement used, some control of emerged weeds and/or incorporation of herbicide in the row area is possible. However, tillage also incorporates some weed seed, rather than removing it from the row.

One modified ridge-till system uses a rotary tiller with planting units mounted behind it. To maintain the old ridge and avoid excessive power requirements, operate the rotary tiller only in the top 2"-3" of the ridge. Limit rotor tines to the row area to provide a strip till configuration. Other ridge-till systems use mulch treaders, rolling stalk choppers or flexible harrows.

*Major reasons for using one of these implements include:*

- Rounding or flattening the ridge.
- Removing some of the residue from the ridge.
- Killing emerged weeds.

Regardless of the tillage implement used, keep operating depth very shallow. Plant stubble in the old row should remain visible. Chemical incorporation is not a goal.

## FALL STRIP-TILL

While the ridges in the ridge-till system worked quite well to provide drainage on poorly drained soils, many producers prefer a strip tillage operation rather than ridges to aid soil warming. This zone tillage operation removes the residue from the row area, allowing sunlight to hit the soil surface and warm the soil. As with ridge planting, planting with strip-till takes place in the residue free strips but usually without the ridge. The adoption of no-till drilling soybeans in narrow rows made ridges unpopular in many areas.

Early strip tillage attachments were typically placed directly on the planter. One, two, or often three fluted coulters tilled and loosened the soil, burying some residue and working air into the soil. This aided in soil drying and warm-up but usually pulled up wet, sticky soil which interfered with the planter units.

To reduce this problem and to allow soil warming to start sooner, some producers placed the coulters on a separate toolbar and performed the strip tillage operation anytime from one-half day to several weeks ahead of planting. Spiked-wheel residue movers and fertilizer knives have also been used to remove the residue and loosen the soil in these zone tillage systems.

While many variations of strip-till have evolved, the most widely practiced strip-till systems perform the tillage in the fall and usually place nutrients in the tilled zone at the same time. A subsoiler, fertilizer knife, or multiple coulters perform the tillage allowing fertilizers to be injected through the knife or behind the coulters. Often, some type of soil mounding disks, such as anhydrous knife slot closing disks, build up a slight ridge in the tilled zone to aid in soil drainage. In drilled soybean residue, producers equip their strip tillage implement with row markers to properly lay out the cleanly tilled strips for planting next spring.

Some do not consider fall strip-till as a separate tillage system but as a variation of either the no-till or ridge-till systems. The Conservation Technology Information Center's definition of no-till includes strip tillage, provided less than one-third of the total row area is tilled. Some no-tillers may perform the strip tillage operation in the spring. Or as another variation, some ridge-tillers use a strip tillage operation on their ridges to inject nutrients and aid in soil warming.

## NO-TILL

Tillage is essentially eliminated with a notill system. Crop seed is placed in a narrow strip opened with a coulter or disk seed furrow opener. With little or no modification, most planters can be used in notill systems. Common attachments include coulters, stronger down pressure springs and extra weight for better penetration. By disturbing only a narrow slot in the residuecovered soil, excellent erosion control is achieved.

Do not shred standing residue prior to planting. Planters, drills and cultivators perform better when residue is standing and attached to the soil, rather than unattached and lying flat. Notill requires surface application of pre-emergence or postemergence herbicides for weed control. One or two properly timed applications may be necessary.

For later planted crops (*e.g.* grain, sorghum, soybeans) apply an early preplant herbicide at 1/2 or 2/3 rate early in the spring when the probability of rainfall is high to provide maximum herbicide performance. This application often eliminates the need for a burndown herbicide and because weeds are controlled early in the growing season, soil moisture losses are reduced. For full season weed control, use a second lighter pre-emergence herbicide application at planting, a postemergence treatment, or crop cultivation. Equip cultivators to handle residue without clogging.

Notill planting is well suited to many soils. Residue, when uniformly spread, increases water infiltration and reduces soil moisture evaporation. Using notill in poorly drained soils covered with large amounts of residue, delays soil warming and drying in the early spring, which delays germination and emergence. When colder, wetter soils are a concern with early planting dates, use notill planter attachments designed to move residue but not soil away from the row. Fall strip till facilitates warming and drying of the row area and may be a better choice.

# 5

# Water Conservation

## INTRODUCTION

Water conservation is a broad term involving all the measures which aim at arresting maximum amount of run-off water, storing or collecting it in safe, artificially raised structures and recycling the stored water again to the field in times of need. The ultimate aim of water conservation is to produce maximum from the limited availability of water, by using it most efficiently. An efficient system of collecting, storing and reusing run-off water is sometimes also referred to as water harvesting.

From the point of dry land agriculture, water harvesting indicates not only storing rain water in irrigation tanks/ponds but also employing mechanical and cultural manipulations of soil to improve its rainfall absorption and moisture retention capacity. However, the measures adopted are location specific.

To decide on the suitable measures to be taken a preliminary inventory of soil water and other resources, a study of the land use pattern, a dialogue with the target group and a study of socio-political, economic and cultural setting are a must. This implies a scientific approach to water conservation. The need for educating, and organizing people to make them accept necessary reforms for effective implementation of integrated water conservation programmes need not be over-emphasized. The chief agent of soil erosion is water. It is estimated that about 5333 million tonnes of soil (amounting to about 16 tonnes/ha) is eroded from Indian territory. Nearly 29 per cent of this is carried into the sea and nearly 10 per cent is deposited in surface reservoirs resulting in loss of storage capacity of reservoirs and occurrence of floods. Conservation of water virtually reduces the soil erosion. Thus the scope of conservation is very wide, encompassing all the measures of soil conservation, moisture retention and nutrient preservation in the soil.

## NEED OF WATER CONSERVATION

Water is absolutely essential for utilization of all other natural resources including land. Agriculture depends on the availability and supply of water either from rainfall or from surface or underground fresh water resources. Most of the water is received through monsoon rains and is obtained from July to September. Only in some rivers of North India, water is received through the melting of snow in summer.

Because of high intensity of monsoon rains coupled with poor infiltration rates of soils, there is a heavy run-off of water. Run-off occurs when amount of precipitation is more than infiltration of water into the soil. Run-off is very heavy particularly in first rains. As a consequence of heavy run-off, the discharge in rivers increases during rainy period. The experience of last 2-3 decades, shows that every passing year the amount of run-off water discharged into the rivers is in creasing. It indicates that the amount of water which was conserved earlier in the site of rainfall (catchment area) has been reduced.

Our country receives more annual rainfall than the global average of annual rainfall. Therefore, water conservation has to be adopted as a national objective, not only for the sake of water availability but also for protecting the soil which is a precious resource.

## WATER EROSION PROBLEM IN INDIA

Out of the total 328 m ha of geographical area of the country, about 143 m ha is cultivated and about 65 m ha is barren uncultivable waste and cultural wastelands. By and large these lands are more or less devoid of protective vegetation and physical barriers and hence are subjected to very serious water erosion. Nature takes 500 to 1000 years to develop 2.5 cm top soil but gets eroded in a course of a year or two. As a result it becomes the most harmful single factor in the deterioration of productive land. Along with the loss of run-off water a colossal amount of soil is also lost.

Overgrazing, deforestation, faulty cultivation, shifting cultivation and carelessly built roads in catchment areas have led to devastating down streams. These include gullying and floods leading to destruction of farm lands and villages, drop in flow during the dry season and consequent loss of crops, and siltation of reservoirs and canals.

The problem has been further aggravated due to high rate of population growth: both human and livestock, resulting in indiscriminate exploitation of natural sources, for meeting the ever-increasing demand for fodder, fuel, fibre and fertilizers. Thus continuous degradation of production base and imbalance in land - water -plant-man-animal system is leading to ecological imbalance and economic insecurity through severe soil erosion both by water and wind and threat to the quality of our life and cultivation.

Even though no systematic survey for accelerating the extent of water and soil loss in different areas of the country, has been carried out, it. is estimated that about 5334 mt of soil is detached annually from its original place.

The country's rivers carry 2052 mt of this, nearly 1572 mt is carried away by the rivers into the sea every year and 480 mt is being deposited in various reservoirs.

*Soil erosion by water is mainly of four types*:

1. *Sheet erosion:* About 4-10 tonnes/ha/year in red soil, 17-43 tonne/ha/year in black soil and 4-14 tonne/ha/year in alluvial soil.
2. *Gully erosion:* About 33 tonnes/ha/year in ravine regions
3. *Hill side erosion:* More than 80 tonnes/ha/year in landslide, mine spoil areas etc.
4. *Stream bank erosion:* Along with soil, about 2.5 mt of nitrogen, 1.5 mt of phosphorus, 2.2 mt of potash, large amount of organic matter and micronutrients are also lost. The loss of nutrients is valued at about ₹.15000 million per annum. Mohanjodaro and Harappa are the examples of this fact. Glaring evidence of neglect of rainwater can be seen in the denuded hill tops all around and the extensive network of raville in the states of Gujarat, Madhya Pradesh, Uttar Pradesh and Rajasthan. Once a thick forest, Rajasthan is now a desert, Conservation of water is prerequisite to conserve soil which is a natural reservoir of plant nutrients. Taking into consideration the magnitude of problems, there is a great scope to streamline the entire water conservation programme more scientifically on micro watershed basis in cluster instead of sporadic efforts. Watershed management programme to conserve water and soil is a new approach and it is getting popularity very rapidly. The programme involves an appropriate combination of agronomic, mechanical, agrostological and afforestation measures, fitting to soil-climatic and topographical conditions of area and socio-economic conditions of the target group.

## TECHNOLOGY OF WATER CONSERVATION

Water conservation technologies cover all methods of conserving water through increasing water use efficiency, enhancing capacity to retain run-off water, and eliminating water pollution. Water use efficiency largely depends on availability and adoption of water saving devices and willingness of the consumers to reduce their total water consumption volumes. Furthermore, existing rules and regulations such as pricing mechanisms, reduce the total volume used, while economic incentives largely affect the choice and adoption of technology for water conservation technologies.

Water conservation processes can broadly be categorized into pre consumer and post consumer based approaches. Pre consumer based approaches involve increasing the efficiency of water extraction, storage and conveyance. Usually a large amount of water is lost either through evapotranspiration or seepage during transfer from the abstraction point to the point of use. It is estimated that, of the total rainfall in Thailand, about 70 per cent returns to the atmosphere through the process of evaporation and transpiration.

Using technologies that can minimize losses such as reducing evaporative losses from reservoirs, seepage losses from canals and water application losses prior to the water being used for economic purposes, can conserve a vast amount of water. In contrast, post consumer based approaches include the use of marginal quality waters, such as slightly saline water from the sea, untreated groundwater from shallow tubewells and rainwater collected from thatched roofs, for washing, toilet flushing purposes, etc.

In Asia, the agricultural sector consumes more than 75 per cent of the total water withdrawn from all sources. About 60 per cent of this water is lost during conveyance and distribution. Engineering technologies to reduce these losses, and to enhance utilization of water in irrigation schemes, are well studied and established practices which are beyond the scope of this book.

In contrast, agronomic technologies such as efficient and modern methods of irrigation (including drip irrigation, sprinkler irrigation and surge irrigation) are less well-known and may be considered water conservation technologies. These technologies are documented in standard textbooks on agricultural and water resources engineering, and are briefly reviewed herein.

Recycling is also an alternative, post consumer technology for conserving and augmenting water supplies. It involves the reuse of water previously used for one purpose for a particular use in another application, before it reaches a natural waterway or aquifer. By using water several times, farms, urban areas and industries can increase the productivity of each litre of water consumed.

Industries can conserve water by changing production processes from open to closed systems. In many industrial plants it is possible to recycle the cooling water. Some industries process water several times, and treat it at the end of its period of usefulness, prior to discharging the water to a natural water course. Reuse conserves raw water and, at the same time, reduces the volume of wastewater as well as wastewater treatment costs substantially.

Several case studies on water conservation practices in India are described. In terms of particular technologies, there are no specific technologies involved, but rather a way of managing conservation practices in individual households and industries. Some of the technologies typically being practised in the Asian Region also are presented herein.

## DUAL WATER DISTRIBUTION SYSTEM

Water reuse, and the reuse of wastewater in particular, is receiving increasingly wide attention, even though it is often considered to be of marginal quality. Use of treated wastewater through dual water distribution and plumbing systems can provide a secondary source of water for purposes such as irrigating private gardens and toilet flushing.

The provision of waters of lesser quality through a separate distribution system for non-potable purposes from alternative sources of supply can help lower the demand for potable freshwater. Application of this technology is largely a matter of cost, acceptance and practice, as the distribution technology involved is not significantly different in dual distribution systems compared to conventional single distribution systems.

In the Kathmandu Valley, where water scarcity is increasing, conjunctive use of shallow groundwater sources for toilet flushing and washing of clothes, together with potable water supplied through the municipal water supply system for other purposes, is being practised in residential areas. A rower pump (hand pump) is fitted to a 3.8 cm diameter polyvi-nylchloride pipe ranging from 6 m to 15 m in length which is driven to the ground. Water from the groundwater source thus tapped is pumped manually whenever needed.

## EXTENT OF USE

This technology is widely used in the Kathmandu Valley in locations where the groundwater table is within 15 m of the ground surface. It is also popular in locations where the municipal water supply is intermittent. People of low income are increasingly using this technology in preference to the higher cost municipal supply.

## OPERATION AND MAINTENANCE

The operation and maintenance of dual distribution systems is simple, and involves keeping the pipe and the pump clean. Maintenance consists of changing the pump washer once a year or whenever it starts leaking. No additional maintenance of the municipal supply system is required, and no changes in municipal distribution system operation are necessary.

## LEVEL OF INVOLVEMENT

Providing dual distribution systems at the household level requires no external involvement relative to the groundwater sourced portion of the system. The municipal sourced portion of the system is generally constructed and operated by the local governmental unit.

## COSTS

The total cost of the groundwater sourced portion of the system is about $55 for a 10 m deep well.

## EFFECTIVENESS OF THE TECHNOLOGY

This technology has helped alleviate the problem of water scarcity in Kathmandu. For an average household, more than 60 per cent of the annual household water requirements is met by using shallow groundwater, which is of lower quality than the municipal water, for non- potable purposes.

## SUITABILITY

Water supplied from the rower pump can be used for toilets, bathing, gardening, car washing, and similar purposes. The dual distribution system is most suitable for use in areas where the groundwater is within 15 m of the ground surface; otherwise, a mechanical pump will be necessary, adding to the cost of the groundwater sourced portion of the system.

## ADVANTAGES

This technology is inexpensive and can be constructed using locally-available technology. Water is made available whenever needed, and use of the dual sourced system eases the water scarcity problem not only at the household level but also throughout the entire city.

## DISADVANTAGES

Conjunctive use of dual sourced water may be limited as a result of poor water quality. Water drawn from alternative sources may not be used for drinking even after boiling because of odours and tastes associated with groundwater. Further, such water may pose severe health hazards if the abstraction point is located too close to septic tank outflows. The possible high nitrate concentrations and bacterial levels that may be present in surfacial groundwaters may also lead to health hazards when used by poor people and children for drinking purposes. Widespread use may contribute to a decline in the groundwater table (due to over exploitation). Also, water logging and mosquito breeding in the pump area may occur if proper drainage is not provided.

## CULTURAL ACCEPTABILITY

No cultural problems have been noted, although use may be limited due to odours associated with the groundwater.

## FURTHER DEVELOPMENT OF THE TECHNOLOGY

This technology will be more attractive and useful if simple and inexpensive electric motors are attached to the tubewell and the groundwater sourced portion of the system is fully incorporated into the household water supply system.

## EVAPORATION REDUCTION

## WATER EVAPORATION RETARDANTS

In India, the use of Water Evaporation Retardants (WER) is being investigated as a water conservation measure in surface waters. Control of evaporative losses is being effected using Ceto-Stearyl alcohol, a blend of saturated fatty alcohols, previously imported at high cost and not readily available in India. India's first fatty alcohol plant was commissioned in 1981 at Jalgaon, Maharashtra, by Aegis Chemical Industries Ltd., using an exclusive technology based on an high pressure hydrogenation technology developed in collaboration with Haldor Topsoe, Denmark. The plant is one of the few in the world and the only one in India, producing alcohols to international specifications. In 1983, Aegis successfully developed an effective water evaporation retardant, Acilol TA 1618 WER, an emulsion based on fatty alcohols manufactured from natural vegetable oils.

Acilol TA 1618 WER was developed in the laboratory after extensive research aimed at identifying a product compatible with the climatic conditions of the country. Field trials were conducted in March 1984 with the help of Gujarat Engineering Research Institute. These trials confirmed the effectiveness of Acilol TA 1618 WER.

A dispensing technique for applying Acilol TA 1618 WER to water surfaces, consisting of barrel tanks with a drip feed arrangement mounted on "Floating Rafts" anchored at different points on the lakes or reservoirs, was also developed to suit Indian conditions. The first major project using this technique to conserve water was undertaken by the Public Health Engineering Department (PHED), Jaipur, at Ramgarh Lake. About 50 mg/m2/d of the chemical is required and can result in a savings of about 30 per cent of the daily loss of water due to evaporation.

## TECHNICAL DESCRIPTION

Evaporative losses can be controlled using various technologies. For example, the use of mono molecular organic surface films has been shown to be an efficient technology for reducing such losses. The mono molecular film is applied to an open surface water storage area and allowed to over the water surface. Typical mono molecular films are comprised of long-chain fatty alcohols such as cetyl alcohol (hexadecanol) and stearyl alcohol (octadecanol).

These chemical not only suppress evaporation but also prevent mosquito breeding in the water. The use of stearyl alcohol in doses of up to 70 g/cm2 can reduce evaporative losses by up to 55 per cent of the loss due to evaporation from a free water surface. However, the economics and extent of use of this technology are yet to be explored and established.

## EXTENT OF USE

Use of surface films to reduce evaporative losses is principally in arid and semi-arid regions.

## OPERATION AND MAINTENANCE

The operation and maintenance requirements of this technology are negligible. Maintenance is required only for the rafts and boats, which may be locally constructed, used during the application of the evaporation retardants. No specialized skill is required.

## LEVEL OF INVOLVEMENT

The implementation of this technology is generally carried out by government departments, primarily the public health engineering and water supply departments. In India, it is now being promoted by various state government agencies and local authorities. Generally, control of evaporation is focused at the government level having prime responsibility for water resources management.

## COSTS

No cost data were available as this technology remains largely experimental.

## EFFECTIVENESS OF THE TECHNOLOGY

Of the various substances capable of forming mono molecular layers on a water surface, fatty alcohols in their pure form have been found to be most suitable and effective in retarding evaporation with no side effects. Savings from the prevention of water loss due to evaporation have been reported to be as high as 0.70 million cubic metres of water (equivalent to one month's water supply for Jaipur City, India) using Cetyl and Stearyl Alcohol as an evaporation retardant.

## SUITABILITY

The technology is most suitable in small surface water storages where there are no strong winds to disrupt the retardant layer.

## ADVANTAGES

Use of evaporation control techniques requires a small capital investment. Locally constructed rafts can be used in its application, and skilled labour is not required, except for operation of motorized boats.

Fatty alcohols present no hazards in handling, and are non-inflammable, non-toxic, and non-irritating. There are no known harmful effects, making this technology safe for application on drinking water lakes and reservoirs. There are also no known ecological ramifications, as the alcohols contain straight chain carbon compounds which are biodegradable and permeable to oxygen.

## DISADVANTAGES

The fatty alcohols used as WERs are not readily available and are costly. Historically, fatty alcohols were produced only by sperm whales or by sodium reduction of animal oils and fats. It was only after the development of high pressure hydrogenation process that good quality fatty alcohols are now commercially available. However, this is a very advanced technology, and requires trained and skilled staff to operate.

## CULTURAL ACCEPTABILITY

There are no known problems with cultural acceptability.

## FURTHER DEVELOPMENT OF THE TECHNOLOGY

More research is necessary in the area of stabilising the chemical film in the face of high wind velocities. Properties such as rate of spreading, specific resistance to evaporation and surface viscosity are required to be measured in evaluating the efficiency of the retardant.

# COCONUT PICK-UPS

## TECHNICAL DESCRIPTION

In South India, coconut pick-ups is a name popularly used for weirs constructed exclusively to provide water to coconut gardens. These weirs or pick-ups are small structures built across the seasonal or perennial streams to slow the flow of water at an appropriate location. This results in surface water storage, groundwater recharge, reduction of soil erosion and availability of water for other purposes. The bunds that form the puck-ips are constructed of locally available materials such as stones, boulders or mud turfed with grass. The bunds are usually built within the water course almost to the height of the surrounding ground level, depending upon the width and steepness of the stream.

When stream flows occur, the coconut plantation, situated within the floodplain on either side of the water course, is temporarily flooded. The water impounded by the structures recedes within 3 to 4 days, with the excess water frequently diverted into a tank or cistern. Sometimes there will be several coconut pick-ups within a catchment, and several tanks which may flow from one tank to the other.

## EXTENT OF USE

This technology has been used in many places in South India.

## OPERATION AND MAINTENANCE

The pick-up is constructed by the farmers and hence both operation and maintenance is the responsibility of the user. The principle maintenance requirement is keeping the pick-up sealed to prevent loss due to leakage.

## LEVEL OF INVOLVEMENT

Usually, the pick-ups are managed by the users who build them. Repair and maintenance are also carried out by the user. Government may be involved in the funding of the pick-ups, and in the initial permitting of the project site.

This function is commonly performed by the local self-governing body that functions as a farmers/users association and manage their activities.

**COSTS**

Construction of a typical coconut pick-up for an area of 10 ha in India would cost about $2 000.

**EFFECTIVENESS OF THE TECHNOLOGY**

The technology is successful in providing adequate water to many coconut farmers for agricultural purposes.

**SUITABILITY**

The technology is suitable for use in tropical areas with moderately rolling topography and small streams.

**ADVANTAGES**

The use of pick-ups encourages groundwater recharge by promoting infiltration and vertical percolation. This recharge helps to sustain yields to percolation tanks and tube wells supplying open tanks. The recharge of open wells is almost immediate, even at a distance of up to 0.5 km depending on soil structure and gradients.

Ponds created by the pick-ups can serve also as a drinking water source for livestock from up to 10 villages. Further, the floodwaters deposit a few millimetres of silt behind the pick-up structure and enrich the soils of the floodplains behind the pick-up with nutrients associated with materials like manure, leaves and other terrestrial debris carried in the run-off flows. These minerals and nutrients enrich the soil particularly during the monsoon season without the use of fertilizer supplements.

**DISADVANTAGES**

The presence of the temporary ponds behind the pick-ups can promote mosquito growth and exacerbate human health risks.

## RAIN WATER HARVESTING

The principle of collecting and using precipitation from a catchments surface. An old technology is gaining popularity in a new way. Rain water harvesting is enjoying a renaissance of sorts in the world, but it traces its history to biblical times.

Extensive rain water harvesting apparatus existed 4000 years ago in the Palestine and Greece. In ancient Rome, residences were built with individual cisterns and paved courtyards to capture rain water to augment water from city's aqueducts. As early as the third millennium BC, farming communities in Baluchistan and Kutch impounded rain water and used it for irrigation dams.

## ARTIFICAL RECHARGE TO GROUND WATER

Artificial recharge to ground water is a process by which the ground water reservoir is augmented at a rate exceeding that obtaining under natural conditions or replenishment. Any man-made scheme or facility that adds water to an aquifer may be considered to be an artificial recharge system.

## WHY RAIN WATER HARVESTING

*Rain water harvesting is essential because*:

- Surface water is inadequate to meet our demand and we have to depend on ground water.
- Due to rapid urbanization, infiltration of rain water into the sub-soil has decreased drastically and recharging of ground water has diminished.

As you read this guide, seriously consider conserving water by harvesting and managing this natural resource by artificially recharging the system. The examples covering several dozen installations successfully operating in India constructed and maintained by CGWB, provide an excellent snapshot of current systems.

## RAIN WATER HARVESTING TECHNIQUES

There are two main techniques of rain water harvestings.

- Storage of rainwater on surface for future use.
- Recharge to ground water.

The storage of rain water on surface is a traditional techniques and structures used were underground tanks, ponds, check dams, weirs etc. Recharge to ground water is a new concept of rain water harvesting and the structures generally used are:

- *Pits*: Recharge pits are constructed for recharging the shallow aquifer. These are constructed 1 to 2 m, wide and to 3 m. deep which are back filled with boulders, gravels, coarse sand.
- *Trenches*: These are constructed when the permeable stram is available at shallow depth. Trench may be 0.5 to 1 m. wide, 1 to 1.5m. deep and 10 to 20 m. long depending up availability of water. These are back filled with filter. materials.
- *Dug wells*: Existing dug wells may be utilised as recharge structure and water should pass through filter media before putting into dug well.
- *Hand pumps*: The existing hand pumps may be used for recharging the shallow/deep aquifers, if the availability of water is limited. Water should pass through filter media before diverting it into hand pumps.

- *Recharge wells*: Recharge wells of 100 to 300 mm. diameter are generally constructed for recharging the deeper aquifers and water is passed through filter media to avoid choking of recharge wells.
- *Recharge Shafts*: For recharging the shallow aquifer which are located below clayey surface, recharge shafts of 0.5 to 3 m. diameter and 10 to 15 m. deep are constructed and back filled with boulders, gravels and coarse sand.
- *Lateral shafts with bore wells*: For recharging the upper as well as deeper aquifers lateral shafts of 1.5 to 2 m. wide and 10 to 30 m. long depending upon availability of water with one or two bore wells are constructed. The lateral shafts is back filled with boulders, gravels and coarse sand.

*Spreading techniques*: When permeable strata starts from top then this technique is used. Spread the water in streams/Nalas by making check dams, nala bunds, cement plugs, gabion structures or a percolation pond may be constructed.

## DIVERSION OF RUN OFF INTO EXISTING SURFACE WATER BODIES

Construction activity in and around the city is resulting in the drying up of water bodies and reclamation of these tanks for conversion into plots for houses. Free flow of storm run off into these tanks and water bodies must be ensured. The storm run off may be diverted into the nearest tanks or depression, which will create additional recharge.

## URBANISATION EFFECTS ON GROUNDWATER HYDROLOGY

- Increase in water demand
- More dependence on ground water use
- Over exploitation of ground water
- Increase in run-off, decline in well yields and fall in water levels
- Reduction in open soil surface area
- Reduction in infiltration and deterioration in water quality

## METHODS OF ARTIFICIAL RECHARGE IN URBAN AREAS

- Water spreading
- Recharge through pits, trenches, wells, shafts
- Rooftop collection of rainwater
- Roadtop collection of rainwater
- Induced recharge from surface water bodies.

# TRADITIONAL METHODS OF SOIL AND WATER CONSERVATION

"Coconut pick-ups" is the term popularly used to describe diversion weirs constructed exclusively to benefit coconut gardens. The "pick-ups" are small structures built across seasonal or perennial streams to slow the flow of water at appropriate locations. This results in surface water storage, groundwater recharge, reduction of soil loss due to erosion, and provision of water for other activities. The technical concept is simple; *i.e.*, to slow the flow of water at strategic places in order to collect water using bunds constructed of locally available materials, like stones, boulders or clay (mud).

Mud bunds are turfed with a specific variety of locally available grass, Maane hullu. The bunds are usually built within an incised stream bed almost to the height of ground level, depending upon the width and steepness of the stream. During periods of peak stream flows, the berm creates an hydraulic obstruction which backs up the water, flooding the coconut plantations located on either side of the stream.

The water impounded in this manner generally recedes within 3 to 4 days. In addition to supplying the plantations with water, the flood leaves a few millimetres of silt, and associated major and micro nutrients and organic matter derived from manure, leaves and the mineral carried downstream, particularly during the early monsoonal flows. Excess water subsequently flows into a storage tank.

Sometimes there will be several coconut pick-ups and tanks within a catchment, and excess flows may be conveyed from one tank to another. This case study illustrates the utility of this technology by reference to a pick-up located in the Tumkur District of Karnataka. Tumkur District receives an average annual rainfall of 688.4 mm.

The net irrigated farming area accounts for 13.8 per cent of the net farming area, and extends over 77 673 ha., 361.7 ha of which are irrigated with water supplied from tanks. Tumkur District stands third in terms of the number of pick-ups in active use, after the Mandya and Chikmagalur Districts.

The Coconut Development Board has provided $150 000 to state governments for pick-up construction under the million wells scheme. The amount spent on each pick-up will vary depending on the command area served by the pick-ups. Normally petitions are submitted by the villagers through their local government bodies (such as gram or mandal panchayats) to the zilla parishad or public works department. A survey of the site proposed by the villagers is conducted by the public works engineer, and an estimate of the construction cost is sent for approval to the district level engineer. After the technical evaluation, a financial sanction is obtained, and a contract is awarded to the registered contractors. Construction work is undertaken usually during the summer season.

## TECHNICAL DESCRIPTION

The construction site is selected in an area with exposed bedrock to ensure a strong foundation. The stream is also narrow at such sites, and the sites typically form convenient sites for road crossings. Upstream of these constrictions, there is generally an area available for spreading the water more widely (the floodplain of the stream), and the back water created by the pick-ups can extend up to nearly one-half of a kilometre in length.

During the construction phase, a cofferdam created by about 200 polyethylene sandbags are used to make a temporary bund. Within this bund, boulders are placed upon the bedrock in a slanting manner to form a permanent bund. Each stone weighs between 100 kg and 1 000 kg, and are generally transported to the site in a power tiller trailer over a distance of between 0.5 km and 5 km. Wooden planks are used to load and position heavy stones.

Behind this rock wall, a supporting vertical wall with a width of 10 cm is built, and both walls are raised simultaneously. Between the walls, an impervious core is created by a layer of "crude jelly" (bitumen) encased within a layer of clay. The clay is wetted to produce a slurry which is compacted by repeated passes of the power tiller to ensure a good seal, and further clay is added until the entire structure is encased in earth. Construction usually takes place over a period of several years.

During the first year, the total height of the bund is only about 5 cm above the stream bed, although the side embankments or wing walls are constructed on either side to a height of 10 cm to prevent the failure of the banks during periods of high flow, when the structure is overtopped. On either side of the bund, stone pillars are erected to enable people to locate the bund during peak flow periods, and to assist them in crossing the stream when the pick-up overflows. During the second year, after observing the water flow of the previous year to determine the effects of any seepage, overflows (possible flanking of the wing walls) and impoundment (to determine the ability of the structure to withstand the water load), further construction occurs and the height is increased by another 4 cm.

The length of the pick-up increases as the height increases. During the second year also, work is completed prior to the monsoon. During the third year, after observing the water flow, the height is further increased by another 4.5 cm until it almost reaches the ground level. On completion, the surface of the pick-up is covered with stone slabs to prevent erosion of the clay surface. At this juncture, the entire structure is encased within a mud embankment, built to a height of 0.75 m (2-1/2 feet) above the stream bed, and turfed with Maane hullu grass to prevent erosion and stabilize the stream banks.

The Tumkur pick-up, constructed in the above manner, is almost impervious and has withstood even the heavy stream flows in the last five years.

## EXTENT OF USE

The Tumkur pick-up is a typical coconut pick-up, and is one of five such structures situated on this water course. The Tumkur pick-up was selected for this study as information was available for th entire period of the construction project and for the 7 year operational period.

It is the third pick-up in the catchment. The watershed of the Huvinahalla Stream which feeds the pick-up originates in the Handanahalli Hills. Stream flow is seasonal, with the stream flow dependent upon the annual rainfall. Nearly 90 per cent of the stream reach passes through coconut plantations or gardens.

The Tumkur pick-up was designed and built by an individual farmer, and constructed using locally available materials. The design was approved by, and construction inspected by, the public works department prior to the implementation of the project. The formal inspection process and granting of operating permission by the public works department is intended to avoid litigation by downstream farmers. The project was completed over a span of three summers, with the height of the pick-up being raised each year until the crest of the weir was almost at ground level. The crest of the weir is stabilized and is used as a cross-over road to reach the other stream bank during high flow periods.

The of the number of pick-ups, and their command areas and beneficiaries in the Tumkur District for the period 1991-92 to 1994-95.

## OPERATION AND MAINTENANCE

The pick-ups are constructed by individual farmers, who are responsible for both the operation and maintenance of the structures. The primary maintenance requirement is that the structure remain impervious to minimize the possibility of dam failure.

## LEVEL OF INVOLVEMENT

The pick-ups are managed by the individual farmers who build, operate and maintain the structure. Government may be involved in the funding and/or initial sanctioning of the project site, and in the inspection of the dam. Mainly, governmental actions will be carried out by the local self-governing bodies involved.

The rights of the farmer who has constructed the pick up are as follows: the farmer may make use of the silt accumulated behind the bund; the farmer may construct a storage tank and provide such tanks with a sump from which to pump water that has been conveyed to the tank from the pick-up (however, neither the farmer nor any other individual can directly pump water from the pickup); and, the farmer can raise bamboo, teak and other suitable species of tree on their katha bund (wingwalls) and along their periphery.

## EFFECTIVENESS OF THE TECHNOLOGY

This technology augments available water resources by storing surface water and recharging groundwater. There was an observed recharge in six open wells as a result of infiltration from the coconut pick-ups from the second year onward. This traditional technology is ideally suited for the local conditions prevailing in this region of India. In terms of costs, the pick-up is valued at about$7 000, based upon the nominal rates used by the government ($353/ha). However, by using locally available materials and using existing farm labour, the farmer actually spent $1 825 in constructing the Tumkur pick-up. It is therefore important to create awareness of this technology amongst the farmers, and encourage local initiatives in its construction.

## ADVANTAGES

*The principle benefits which may be derived from using this technology include*:

- The availability of impounded water, the depth and extent of which depends upon the terrain, width and depth of the stream, and the duration of which depends upon the soil type, usage and frequency of rainfall/refilling.
- Enhanced groundwater recharge due to both vertical percolation and horizontal interflow, revitalizing percolation tanks, open tanks and tube wells; the recharging of open wells is almost immediate within a radius of up to one-half of a kilometre depending on soil structure and gradients.

The ponds created by the pick-ups serve as drinking water sources for livestock from 8 to 10 villages in the communities surrounding the pick-up. These ponds also serve as a common place for washing clothes and conducting religious ceremonies (*Ganga Pooja*); for seasoning wood and wooden poles by submerging them in water, which is a common practice; for retting coconut husks and agave leaves; and for other, similar activities. The waterbodies attract birds and serves as habitat for aquatic creatures, and can provide water for pisciculture. Agroforestry practices can be implemented along the periphery of the waterbody, which creates a microclimate that is more congenial for the growth of plants, including coconut and arecanut plants.

The pick-ups reduce the impacts of soil erosion and downstream siltation, while retaining silt that can be collected in the pick-ups and carted off to the coconut plantations during summer months to improve soil structure and fertility, and, in turn, increase their productivity. The pick-up structures can also serve as a vital link between villages, especially during the rainy season when travel may be restricted by flooding rivers. In certain areas, seepage into percolation wells as a result of groundwater recharge from the pick-ups is used to irrigate coconut and arecanut gardens, other field crops like ragi and maize, and even paddy crops.

## DISADVANTAGES

Possible disadvantages include increased nuisance due to mosquitoes breeding in stagnant water that may collect upstream of the pick-up structures.

## FURTHER DEVELOPMENT OF THE TECHNOLOGY

Based on the aforementioned observations, methods of involving the local people in planning and executing such projects is very important to achieving long lasting benefits from such ventures.

## RECYCLING AND WATER CONSERVATION

## GUJARAT STATE FERTILIZER CORPORATION, INDIA

Gujarat State Fertilizer Corporation (GSFC) is one of the largest integrated fertilizer and petrochemical complexes in India, producing a variety of fertilizers, intermediates and petrochemical products. GSFC has adopted an integrated approach to conserving water.

This strategy has brought multiple benefits to the operations of company. By recycling their effluent streams, GSFC substitutes recycled water for raw water in their water stream, resulting not only in water conservation and cost savings, but also in the recovery of chemicals previously discharged in the process and an higher level of water pollution control compliance. Water consumption has been maintained at a low level, despite the expansion of the plant and increased production levels.

*The raw water supply to GSFC is met from two sources*:

- From a joint water supply scheme with Gujarat Refinery, using French-type, radial collection wells situated in the bed of the Mahi River, which provides up to 36 370 m3/d.
- From GSFC-owned French-type, radial collection wells, situated in the bed of the Mahi River at Parthampura, which supply up to 45 460 m3/d.

The actual throughput of these wells is dependent on groundwater levels. During drought periods and in the summer months, the groundwater levels drop, limiting the throughput of the wells. Nevertheless, the primary source of water supply to the GSFC operations is the jointly-operated Gujarat Refinery well and balance is met from the GSFC-Parthampur installation.

The total daily water requirement of the GSFC operation prior to the installation of the water conservation and recycling practices as about 45 000 m3/d. Specifically, the water requirement of the GSFC plant, after Phase I, II and III expansions in 1977, was:

| | |
|---|---|
| Cooling Tower Make-up Water | 24 400 $m^3$/d |
| Demineralized Water Production | 9 000 $m^3$/d |
| Process Water | 3 000 $m^3$/d |
| Fire Protection Water | 1 000 $m_3$/d |
| Drinking Water | 3 000 $m_3$/d |
| Township Water Supply | 5 000 $m^3$/d |
| Total | 45 400 $m_3$/d |

With the implementation of the integrated approach to water conservation and recycling of effluents, the present water requirement of the GSFC complex is 40 000 m3/d. Use of these technologies has helped to maintain water demands at GSFC at a low level, despite an increased level or production and an increased number of operating divisions.

## TECHNICAL DESCRIPTION

GSFC opted for an integrated approach to water conservation and recycling based upon the philosophy that conserving water conserves all resources associated with the water. Conservation of steam, condensate, demineralised water, and process water leads to the conservation of water with maximum returns.

For example, within a network of plants, it was possible to recycle waste stream from one plant to another plant. As a practical result of this recycling philosophy, the phosphatic group of plants achieved the total recycling of its effluent, conserving water, recovering previously lost product and controlling pollution. Similar strategies were adopted in the ammonia/urea group of plants. Some of the actions taken to conserve water are elaborated in the following sections.

## RECYCLING ACIDIC EFFLUENTS IN CHALK PONDS

Chalk is a by-product produced by the ammonium sulphate plant. Chalk slurry is pumped to chalk ponds where it is mixed with highly acidic, phosphoric acid contaminated return flows. The acidic effluent is neutralized by chalk slurry and the chalk floc settles in the pond.

Two chalk ponds have been sealed with polyethylene linings on their bottoms to minimise water percolation. After a period of operation, the ponds fill with chalk and must be emptied; hence, the requirement for two ponds to ensure continuous operation of the plant.

During the time when the first pond is off-line and being emptied, the second, empty chalk pond is filled with water to bring it on line. Annually, 170 000 m3 to 180 000 m3 of chalk is reclaimed using this process and an equivalent amount of water is consumed in filling the chalk ponds before they are commissioned. Cooling water can be used to meet this initial water requirement.

Alternatively, effluents from ammonia/urea, melamine, and caprolactum plants can be used after treatment to strip the ammoniacal nitrogen. These effluents are collected in a central collection pond, pumped to the polyethylene-lined ammoniacal effluent lagoon, and treated in an Air Stripping Tower to remove ammoniacal nitrogen, before being discharged to the chalk pond or disposed. In normal plant operations, this reclaimed water is also used as make up water for the chalk ponds.

## RECYCLING BAROMETRIC CONDENSER WATER AS COOLING WATER

In the evaporation section of the ammonium sulphate plant, there is a surface condenser followed by two barometric condensers, for vacuum generation. The gases, after coalescing in the surface condenser, are condensed in the barometric condenser through direct contact with the cooling water. The barometric condensers use cooling water at a rate of 135 m3/hr. Rather than discharge this cooling water, as was previously the case, the barometric condenser water is now segregated from the main effluent disposal grid and is pumped back to cooling tower of ammonium sulphate plant, recycling 135 $m^3$/hr of cooling water.

## RECYCLING CONTAMINATED CONDENSATE FOR CHALK REPULPING

Process water condensate, generated in the evaporation process employed in the ammonium sulphate plant, is contaminated and cannot be directly reused. Thus, in excess of 30 m3/hr of process condensate had been historically discharged as effluent. However, water was required elsewhere in the ammonium sulphate plant to repulp chalk after it had been used in the filtration of ammonium sulphate plant liquor.

The filter cloth must also be washed with water at the same time. In order to conserve water, a system was designed to substitute process effluent, mainly process water condensate, for the non-recycled cooling water that had been previously used for this purpose. All contaminated process water condensates from different sources with GSFC are collected in a central collection pit, and pumped to the chalk filter to wash the filter cloth and to repulp the chalk. The resultant chalk slurry is pumped to the chalk pond for settling and neutralization of acidic wastewaters as described above.

It should be noted that, as this repulping and washing stage is one of the most critical in the entire operation, automated safeguards were provided to ensure that the process remained unaffected in case of any problems being experienced with the effluent recycling system. However, the recycling system is working well, with two ammonium sulphate plants being successfully operated with total recycling of the effluents to the chalk pond. This has provided significant savings in cooling water requirements, ehanced recovery of ammonium sulphate, and increased the level of water pollution control achieved.

## RECOVERY OF PURE CONDENSATE AS BRINE-FREE WATER (BFW)

In the ammonium sulphate plant, a 40 per cent ammonium salt solution is evaporated to produce ammonium sulphate crystals. Steam is used as the heating medium to evaporate the saline solution. Condensate from both parts of the process, previously lost as waste, is now captured in the main condensate grid as part of the process design. The purity of the condensate is analysed, with the pure condensate being directed into a separate circulation system and pumped to the steam generation plant to be used as brine-free water. Quality safeguards and process safeguards are provided so that process is not affected by any malfunction of the operational control systems.

## RECYCLING PHOSPHORIC ACID PLANT EFFLUENT FROM THE CHALK PONDS

In the phosphoric acid plant, water from the chalk ponds, described above, is utilized in the fume scrubbers, condensers, and flash cooling systems, and in other, miscellaneous services. The acidic return flows from the plant are pumped back to the chalk ponds for neutralization, settling and natural cooling.

The cooled chalk pond water is returned to the phosphoric acid plant and remains in circulation until the build up of dissolved solids (TDS) in recirculating water begins to impair its effectiveness as a coolant, at which point, the chalk pond water is bled off to control the TDS build up. The water lost through this bleed off is subsequently made up by the addition of new cooling water. The high TDS water is discharged as effluent.

Also in the phosphoric acid plant, process water was used for gypsum repulping and washing of the cloth pan/belt filter. The cloth filter captures the gypsum cake and conveys it to a discharge point, after which the cloth is washed by a number of spray nozzles located on both sides of the belt. To conserve water, the same wash water is used for repulping the gypsum cake and conveying the gypsum slurry to the drum filters where it is further purified.

As a further conservation measure, chalk pond water is substituted for process water throughout the process. Chalk pond water is mixed with condensate for use in filter cloth washing and gypsum repulping. As a further benefit of this recycling scheme, about 2 500 metric tonnes of ammonium sulphate is recovered annually from the chalk pond.

The chalk pond water contains 1.5 per cent to 2 per cent ammonium sulphate which is recovered in the phosphoric acid process in form of diammonium phosphate (DAP). The recovery of ammonium in the form of DAP has had a tremendous impact on profitability, and has clearly demonstrated the benefit of the integrated effluent recycling, recovery and pollution control programme.

## RECYCLING CHALK POND WATER IN THE GRINDING MILL DUST SCRUBBER

In the phosphoric acid plant grinding mill, there is dust scrubber to recover rock phosphate dust downstream of the product cyclones. This is a wet scrubber, which used process water at a rate of 95 m3/hr. The resultant slurry was recycled to the phosphoric acid digester. As part of the water conservation programme, chalk pond water was substituted in place of the process water. Recycling process water condensate in the gypsum purification section In the gypsum purification section of the phosphoric acid plant, gypsum cake was collected on a drum filter and washed off the filter using hot water jets. Process water, heated with live steam, was used for this purpose at a rate of 5 m3/hr. As part of the integrated water conservation programme, hot condensate was used in this process.

## DEMINERALISED WATER CONVERSION IN BAROMETRIC CONDENSER COOLING TOWER

In the urea plants, cooling water is used in the crystallization section of the barometric condensers for vacuum generation. The barometric condensers are cooled in the barometric condenser cooling tower.

The cooling water used in this process is contaminated with ammonia and urea due to its closed loop circulation and to process upsets which result in carry overs of ammonia and urea. Hence, it is necessary to make up the cooling water supply by bleeding the contaminated cooling water from the cooling tower to maintain water quality.

This created a continuous flow of liquid ammoniacal effluent from the bleed of the cooling tower, and the subsequent consumption of cooling water supplies. As part of the water conservation programme, the cooling towers were converted to a demineralized water circulation and cooling water system in which the cooling towers were isolated. After demineralized water circulation, excess water from this cooling tower was substituted for demineralized water used in the plant.

## RECYCLING OF EFFLUENT IN THE DIAMMONIUM PHOSPHATE PLANT

In the diammonium phosphate plant, pre-neutralizer temperature control is accomplished with the addition of process water. Temperature control is quite critical for plant operation. As part of the water conservation programme, water from washings and leakages, etc., was collected in a pit and substituted for process water in the temperature control function. As in other critical systems, an automated temperature control system was retained in the process so that the process is not affected by problems with the effluent circulation system.

## RENOVATING WATER AT THE SEWAGE TREATMENT PLANT

GSFC Township houses about 1 700 families in a setting that has vast areas of open land with lawns, plants and recreational facilities. Sewage water from the Township was pumped into the main effluent grid and discharged with other plant effluents at a rate of 230 m3/hr. As part of the water conservation programme, an activated sludge sewage treatment plant has been installed and commissioned recently, which reclaims about 135 m3/hr of treated effluent as cooling water and irrigation water for the Corporation's experimental farm.

## RECYCLING OF CLEAN WATER IN THE COOLING TOWERS

In the industrial complex, there are number of applications in which cooling water is used for open cooling. The spent cooling water was typically released as waste through the sewers.

To minimize this loss of water, many of the open cooling water applications were converted to jacketed cooling water operations. Cooling jackets were installed on transfer lines, secondary transformers, high tension shift convertors, and other equipment, with the return flows of spent cooling water sources being diverted to cooling towers.

Similarly, water from air conditioner package units was diverted to suitably modified cooling towers. All of these measures resulted in a substantial savings of cooling water.

## OPERATION AND MAINTENANCE

In every water conservation project which recycles cooling water or substitutes recycled water for process water, continuous monitoring is needed to ensure that the recycling practices do not result in any disruptions to the process, or to contamination or problems of corrosion in the equipment.

*In the case of GSFC, special attention is given to*:

- Monitoring of the chalk pond effluent, as well as monitoring of the ponds for any seepage into the ground
- Monitoring of cooling water quality at critical points in the cooling water stream
- Monitoring pump performance, especially of those pumps handling effluent
- Maintaining and testing the automatic safeguards installed to ensure continued plant operations in the case of problems with the effluent recycling system, especially at critical points in the process
- Analysing the purity of the condensates during the recovery of ammonium sulphate crystals, and providing the necessary quality safeguards and process safeguards so that process is not affected by any malfunction in the control systems
- Controlling the continuous circulation of recycled water to maintain optimal cooling effectiveness in the plant, and replacing the recycled water with fresh water as necessary to minimize TDS build up
- Maintaining and testing the automatic temperature control systems to ensure continued plant operations in the case of problems with effluent circulation system, especially at critical points in the process.

## LEVEL OF INVOLVEMENT

The project was implemented at the individual industry level, with major involvement of the senior middle level management of the company. The cooperation of the staff and the support of the top management were the additional factors in its success.

## EFFECTIVENESS OF THE TECHNOLOGY

Various benefits were achieved as a result of the water conservation projects implemented at GSFC. A summary of the quantifiable benefits associated with this integrated programme of water conservation and recycling is presented in Table for each component activity of this project. In addition, there were numerous unquantifiable benefits derived from the project, which are not listed.

## ADVANTAGES

**Table. Water Conservation Benefits Achieved through Integrated Water Management in Industry.**

| Project | Quantified Benefit |
|---|---|
| Recycling of effluent from the chalk ponds | Annual savings of 170 000 m3 to 180 000 m3 of cooling water were attained. |
| Recycling barometric condenser water in cooling water | Recycling of 135 m3/hr of cooling water was achieved. |
| Recycling of contaminated condensate for chalk re-pulping | Savings of 30 m3/hr of process water condensate as well as ammonium sulphate recovery. |
| Recycling chalk pond water in the phosphoric acid plant | Recovery of about 2 500 metric tonnes of ammonium sulphate in the form of diammonium phosphate. |
| Recycling process condensate for gypsum purification | Savings of 5 m3/hr of process water by substituing hot condensate. |
| Renovating water at the sewage treatment plant | About 135 m3/hr of treated sewage is recycled as cooling water. |

The advantages of undertaking water conservation projects were several. First, GSFC conserved water resources while increasing their productivity and profitability. Second, water conservation led to a significant reduction in the cost of water purchased. Third, water conservation reduced the volume of effluent generated in the production process and reduced the cost of effluent handling and treatment. Fourth, the energy required for plant operations was also greatly reduced. The programme also enhanced the ability of the Corporation to achieve its water quality goals.

## DISADVANTAGES

In all of the water recycling projects, and especially in those related to cooling water within the GSFC plants, there was an increase cost associated with monitoring water quality to optimize both the cooling benefits and level of water conservation. Additional control systems were also required to ensure the proper functioning of the plants which use recycled flows.

## FURTHER DEVELOPMENT OF THE TECHNOLOGIES

Water conservation is an attractive option for large and complex industries like GSFC to reduce the water costs, increase production and decrease the consumption of energy. The experience in a public sector organization like GSFC has shown that water conservation is possible and profitable on a large scale. The experience at GSFC can be easily transferred to other fertilizer or other complex industrial units for getting similar benefits.

# WATER EROSION

## NATURE OF THE ISSUES

Water erosion of soils occurs when soil particles are detached and carried away by water flowing across a landscape. In some cases soil loss is uniform (sheet erosion). In other cases small channels are formed (rill erosion). When the velocity and volume of water are high enough, and the soil surface is vulnerable, deep channels can be cut (gully erosion). Tunnel erosion occurs when the subsoil is removed while the surface soil remains relatively intact, producing tunnels under the soil, which eventually cause the surface to collapse (Coles and Moore 2001).

Like wind erosion (Section 6), the on-site impacts of water erosion include soil loss, reduction in soil nutrients and organic matter (including soil organisms), release of soil carbon to the atmosphere, undesirable changes in soil structure, reduced water infiltration and moisture-holding capacity, and exposure of unproductive saline and acid subsoils. Off-site impacts include sedimentation of waterways and impacts on quality of surface water and groundwater (turbidity, nutrient and other chemical loads).

# PRACTICES IN RELATION TO WATER EROSION

Land uses that affect water erosion do so primarily via their effects on ground cover, evaporation of soil moisture, soil structure, compaction by heavy equipment or running of stock, and creation of contours that control water flow (Australian State of the Environment Committee 2011).

## BROADACRE CROPPING

Many of the effects of cultivation on susceptibility to wind erosion (Section 6) also apply to water erosion. Water erosion associated with cropping was recognised as a serious issue in the 1930s (Carey *et al.* 2004). Different studies report sediment yields from cultivated basins of between 2 and 21 times those from undisturbed native forests (Neil and Galloway 1989; Neil and Fogarty 1991; Erskine *et al.* 2002), although it should be noted that good land management can keep these figures within the low end of this range (Erskine *et al.* 2002). Soil conservation structures (contour banks and grassed waterways) were designed to reduce the slope length and thus net water erosion. These have been implemented extensively in Australia, but have not been sufficient to bring soil erosion within acceptable limits.

Management of water erosion on cropping lands has increasingly focused on methods of planting and managing crops and controlling weeds that involve little or no tillage, retention of stubble after harvesting, inclusion of a pasture phase between crops and minimisation of the effects of machinery by controlled traffic methodologies. Creating raised beds for crops in waterlogged areas can create an erosion hazard unless slopes and ground cover are managed carefully.

Over the last 20 years new tillage practices have been developed that maximize water infiltration and reduce run-off; new row spacing and plant arrangement schemes have been developed to reduce soil temperatures and soil evaporation losses. Crop modelling and weather prediction capabilities have been developed to advise farmers on the opportune time of sowing that ensures adequate supply of stored soil water in combination with sufficiently high growing season rainfall probability required to satisfy the crop growth requirements and the farmers' yield goal.

While including a pasture phase between crops is considered advantageous in managing ground cover, the potential effects of stock on the soil surface during this phase can potentially pose similar problems to those faced on dairy farms, especially if soils are wet.

The uptake of minimum tillage approaches has required two major innovations: equipment capable of planting in stubble; and effective methods for weed control without disturbing the soil. The advent of better ways to manage heavy vehicles (controlled traffic) has also contributed to reducing run-off-driven erosion (Li *et al.* 2007).

## HORTICULTURE

As a form of cropping, horticulture faces many of the same risks as broadacre cropping in terms of encouraging soil erosion. The hardening of soils in many orchards (coalescence) restricts the growth and function of tree roots and infiltration of water to roots (Cockcroft 2012). Two key management innovations in orchards have been control of machinery traffic to minimise soil compaction, and establishment of ground cover plants that both minimise erosion and contribute to the soil ecosystem. Increased ground cover is correlated with higher diversity of soil organisms, which has been found to have beneficial effects on water infiltration (and therefore reduced run-off erosion) promotes natural pest control.

## DAIRY

Many dairy farms combine the running of dairy cattle with beef cattle, cropping and/ or irrigated pasture production (Ashwood *et al.* 1993). To maintain high production of milk, pastures are fertilized. Key challenges for such enterprises include controlling sediment (along with nitrogen and phosphorus) losses into waterways, which can be exacerbated by compaction and disturbance of soil by the feet of grazing animals.

Irrigation itself has the capacity to increase soil erosion by accelerating mineral weathering, transporting and leaching soluble and colloidal material, changing soil structure, and raining the local water table, thereby increasing the risk of salinity. Irrigation also has the capacity to reverse soil preparation measures such as the tillage that precedes planting.

## GRAZING

Livestock grazing is the most widespread Australian land use. Impacts of livestock grazing on ground cover. These impacts affect vulnerability of landscapes to both water and wind erosion. In addition, as discussed above, grazing during a pasture phase between cropping could increase vulnerability of soils to water erosion by disrupting soil structure and reducing ground cover.

# 6

# Soil Water Relations

Most vascular plants are rooted in the soil from which they obtain both water and mineral salts. The entrance of water from the environment into any of the organs of a plant is called the absorption, intake, or uptake of water.

In most vascular species the quantity of water absorbed through organs other than the roots is of negligible importance. Any thorough consideration of the entrance of water into roots requires a consideration of the properties of soils, particularly as they affect the movement of water from the soil into the root.

## CONSTITUTION OF THE SOIL

The soil matrix which is the habitat of most roots is an extremely complex system. In general five different components of this system are distinguished:1. The Mineral Matter of the Soil.The parent substance of practically all soils is rock, of which there are numerous varieties. By various weathering processes rock strata are reduced to fragments of diverse sizes and these compose the bulk of most soils. The rock particles in soils may vary in size from stones and gravel down to sub-microscopic particles of colloidal clay.

The mineral portion of the soil is customarily classified into several fractions depending upon the size of the particles. Such classifications usually disregard the very large particles of the soil (rock fragments, pebbles, etc.). The proportions of the different fractions present are very different in different kinds of soils.The clay fraction of the mineral portion of soils requires special consideration.

Unlike the coarser fractions which are composed of small fragments of unmodified rock minerals such as quartz, feldspar, and mica, the clay portion of soils is made up almost entirely of the products of chemical weathering of the rock minerals and hence differs not only in its physical state, but in its chemical composition from the particles in the coarser fractions.

Most of the particles in the clay fraction are of colloidal dimensions and exhibit the characteristic properties of colloidal systems. The particles of the clay complex, like those of many other colloidal systems, retain water by imbibition, *i.e.* within the structure of the particle. On the contrary the sand and silt particles of the soil retain water only on their surfaces.

The presence of a considerable proportion of clay in a soil therefore endows it with a high water retaining capacity. Changes in the water content of colloidal clay result in marked changes in its volume. One result of such changes in volume is the commonly observed cracking of a soil rich in clay upon drying. The plasticity and cohesiveness of soils are also due very largely to the colloidal clay present.

Like other colloidal systems colloidal clay is markedly sensitive to the influence of electrolytes. The micelles of colloidal clay usually bear a negative charge when in contact with water. In the presence of calcium ions the individual clay particles are more or less completely flocculated into compound particles.

Much of the colloidal clay in most soils exists as enveloping films around the larger soil particles and is also closely associated with organic material in the soil. Flocculation of the clay particles therefore usually results in the formation of compound granules including sand and silt particles and organic matter in addition to the clay.

These are called soil crumbs. For agricultural purposes a well developed granular structure of the soil is highly desirable since such a structure favours both a high moisture-retaining capacity and a good aeration of the soil. Calcium soils, that is, those with a high calcium content, are in general the most suitable and most valuable for agricultural purposes partly because calcium favours the development of a crumb structure.

In the presence of an excess of univalent cations ($Na+$, etc.), on the other hand, a greater proportion of the clay fraction of the soil disperses into its ultimate particles, and the soil has a single grain structure. In this condition the soil is in the least desirable structural condition for agricultural purposes. Many soils contain hydrogen ions in excess. Such soils often develop a good crumb structure, but the crumbs are less stable than those developed in a calcium soil. The addition of lime improves the physical structure of such soils. The crumb structure of a soil, especially in its surface layers, can also be destroyed by purely mechanical effects, such as trampling,or beating by heavy rains. The Organic Matter of the Soil.

Most soils contain organic matter which has been derived principally from the partial decomposition of plant residues. Small quantities may also originate from animal residues and excretions. The proportion of organic matter present may vary from almost none as in some sand deposits to 95 per cent or better in some peat soils. In ordinary agricultural soils the amount present seldom exceeds 15 per cent.

In forests organic matter comes from falling leaves, dead branches and trunks of woody plants, roots which die and decay underground, and from dead herbaceous vegetation. In grasslands the underground roots and rhizomes as well as the aerial parts of the plants all contribute their quota to the organic matter of the soil. In well managed agricultural soils attempts are made to maintain the organic matter content by supplying them with organic fertilizers.

The organic matter is the seat of most of the microbiol-ogical processes occurring in the soil. One of the most important of these is the oxidation of the organic matter, a process due largely to the metabolic activities of bacteria and fungi, although a limited amount of purely chemical decomposition probably also occurs. Under conditions which are exceptionally favourable for the activities of microorganisms the organic matter of the soil is oxidised completely and disappears.

For this reason the organic matter of soils in tropical regions, particularly when under cultivation, is very low. Even in more temperate regions cultivation of the soils generally results in a rapid reduction inorganic matter content, due principally to the better aeration induced by till age. As a result of the decay process there is present in most soils organic matter in various stages of decomposition. A large proportion of the organicmaterial which is added to some soils survives in the form of a dark-colouredamorphous substance called humus. Humus is composed principally of the degradation products of the cellulose and lignin derived from plant remains. The accumulation of humus in soils is furthered by conditions unfavourable to the oxidative decomposition of organic matter.

The decomposition of organic matter in bogs, ponds, swamps, and water-logged soils under conditions which are largely if not entirely anaerobic results in the production of relatively large quantities of humus, frequently of the type which is called peat. Where the organic matter supplied is distinctly acid, as under coniferous forests, or heath plants, humus usually accumulates as a definite layer at the surface. Decomposition of the organic remains under these conditions is largely effected by fungi.

In prairie and steppe regions, where a grassland vegetation is predominant, humus usually accumulates in considerable quantities as a result of the decay of both underground and aerial organs of plants. The amount of humus which accumulates in any soil depends upon the relative rates of the addition of organic residues and of its disappearance as a result of oxidation. Soils with a low humus content may result from a sparse contribution of organic matter, as in desert or semi-desert regions, or from a rapid oxidation of organic material which prevents accumulation of humus even when the supply of organic residues to the soil is large.

This latter condition obtains in many tropical and sub-tropical soils. Humus is essentially colloidal in its properties and possesses an even greater imbibitional capacity than the colloidal clay particles of the soil. The plasticity and cohesiveness of the colloidal organic matter, while considerably less than that of the colloidal clay, is much greater than that of the non-colloidal fractions of the soil. Humus is rather inert chemically, its influence on the soil being largely a physical one. Its presence in soils favours a looser structure and hence better aeration, a higher water holding capacity and a reduction in cohesiveness as compared with heavy clay soils.

Since the clay fraction and the humus are the two essentially colloidal fractions of the soil, and are associated in an intimate physical relationship many soil scientists refer to these two soil fractions, considered jointly, as the colloidal complex of the soil. A large part of this colloidal complex occurs in many soils, as films enveloping the larger soil particles. The relation of these films to the maintenance of the crumb structure of the soil has already been mentioned. Many other important soil properties are either due to or greatly influenced by the colloidal complex.

## SOIL WATER AND THE SOIL SOLUTION

Water is universally a component of soils, although the amount present may vary from the merest trace to a quantity sufficient to saturate the soil, *i.e.* completely fill all of the spaces between the soil particles. Dissolved in the soil water are varying quantities of numerous chemical compounds.

These originate principally from the dissolution or chemical weathering of the rock particles, from the decomposition of organic matter, from the activities of microorganisms, and from reactions between the roots of plants and the soil constituents. It is thus more accurate to speak of the soil solution than of the soil water, although in discussions of the water relations of soils it is often a common practice to disregard the presence of solutes in the soil water.

The concentration of the soil solution in any given soil varies with the proportion of water present. While in most soils the soil solution is very dilute, in saline and alkaline soils it maybe so concentrated that only a few species of plants can survive with their roots in contact with it. The principal cations found in the soil solution are $Ca_{11}$, $Mg_{11}$, K1, Na1, $Al^{+1+}$, and $Fe^{111}$ (or $Fe^{++}$); the principal anions are $HCO_3$Th $P0_4$, Cl, $N0_3$, $S0_4$, and 5i03. Other solutes too numerous to mention may also occur in the soil solution but are usually present only in very low concentrations.

## THE SOIL ATMOSPHERE

The irregularity of the soil particles in size shape, and arrangement insures the existence of a certain amount of space between them, even in the most tightly packed soils. This is termed the pore space of a soil. The pore space of soils varies from approximately 30 per cent of the volume of the soil in sandy soils to about so per cent of the volume in clay soils, or even higher in soils which are very rich in organic matter. The pore space of any given soil depends upon the physical and chemical conditions to which the soil is subjected. Conditions favouring a crumb structure of a soil, for example, usually result in an increase in its pore space.

The interstitial spaces of a soil may be occupied entirely by air, as indesiccated soils, entirely by water, as in saturated soils, or, as is most usually true, partly by water and partly by air. The relative proportions of water and air present in any given soil vary depending upon the water content of the soil.

The air present in the soil is generally referred to as the soil atmosphere.The soil air usually contains a higher concentration of carbon dioxide and a lower concentration of oxygen than the above ground atmosphere.

Concentrations of carbon dioxide as high as 5 per cent have been recorded for the soil atmosphere; such values are far in excess of the average value of 0.03 per cent for the air. The accumulation of carbon dioxide in soils is due more to the metabolic activities of micro-organisms than to respiration of soil animals and the underground portions of vascular plants. Except in very dry soils the soil atmosphere is usually saturated or nearly so with water-vapour.5. Soil Organisms.

The soil flora includes bacteria, fungi, and algae. The bacteria are generally the most abundant of all the living organisms present in any soil. Among them are the nitrifying, sulfofying, nitrogen fixing, ammonifying, and cellulose decomposing bacteria.

The bacteria accomplishing the oxidative decomposition of cellulose and similar compounds are the most important agents in the production of humus. The numbers of bacteria present vary greatly from soil to soil, and in any one soil vary with seasonal and other fluctuations in soil conditions. Most soils contain between two million and two hundred million individual bacteria per gram of soil.

The number of bacteria decreases rapidly with increasing depth, sub soils being sterile or practically so. In general an abundant representation of most species of bacteria is favoured in soils by warm temperatures (3 - 400 C.), good aeration, and a good but not superabundant water supply. A high calcium content of the soil is also favourable to the development of most species of bacteria, apparently because it favours a granular structure of the soil, thus improving aeration.

Some species of bacteria, on the other hand, are anaerobic,and thrive when aeration of the soil is deficient. The denitrifying bacteria and certain of the nitrogen-fixing bacteria (Clostridiurn spp.) are examples of anaerobes. Fungi are, in general, most abundant in soils of acid reaction. In such soils they largely replace bacteria as agents of decomposition of organic matter.

The soil fauna includes protozoa, nematodes, earthworms, insects, insect larvae, and burrowing species of the higher animals. The earthworms are generally credited with having the most important effects on soil structure, at least in many soils. Their activities result principally in a general loosening of the soil, which facilitates both aeration and distribution of water. Many of the other soil animals have similar effects on the structural organisation of the soil.

## SOIL PORES AND WATER CHARACTERISTICS

Soils are made up of three parts: mineral materials, organic matter and space (called pore-or void-volume). The relative importance of these varies with soil type but pore space can occupy about half the volume of a medium-textured soil. At optimum water content for plant growth, approximately half the pore space is filled with water and half with air.

The proportions of water and air can change rapidly depending on weather, evapotranspiration and other factors. The dimensions (size, shape and arrangement) and number of pore spaces are most important in determining soil water and soil structure.

Porosity is the volume of soil voids (*pore* space). It is expressed in relation to the bulk volume of the soil. The water holding capacity of a soil depends on its porosity, and the size distribution of its pores. Small pores retain water at greater suctions than larger pores. The moisture (or water) potential is the amount of energy required to remove water from a soil; field capacity is the water-holding capacity after a free-draining soil has been allowed to drain. Available soil water (ASW) is the amount of water which is available for uptake by plants, namely that held at suctions between wilting point and field capacity.

It varies with soil type and can be correlated with the clay content and structural arrangement of the soil. It varies also with soil treatment because the size and distribution of pores in the topsoil reflects surface exposure, normal seasonal wetting and drying, and management. Williams *et al.* (1983), studying the water content of 244 soil samples, found that the ASW of well-structured soils was one-third to twice as large as that in comparable (similarly-textured) poorly structured or degraded soils. Bearing in mind that ASW varies with natural weathering and management,

Hydraulic conductivity (K) of a soil is its conductivity to movement of water down a pressure gradient. High values of K are associated with well-structured soil and contiguous pores; they allow high infiltration rates and rapid drainage. Earthworm channels, which can have populations of 500 $m^{-2}$ in Mediterranean climates, and continuous deep voids left by dead roots (5-10 000 $m^{-2}$) contribute greatly to hydraulic conductivity.

Hydraulic conductivity varies with soil type and management. K values below 10 mm/h are low and likely to cause run-off following rainfall or problems with irrigation, given that steady rain falls at about 10 mm/h. K values of 10 to 20 mm/h can give intermittent run-off (a downpour falls at about 50 mm/h) while values up to 120 mm/h are associated with occasional, increasingly rare run-off. Values above 120 mm/h may facilitate regular drainage to the groundwater, causing potential problems for heavily-fertilised soils, and those treated with effluent, herbicides or pesticides.

Both soil water content and saturated hydraulic conductivity generally relate to the number and continuity of pores, particularly the larger macropores. It is, however, difficult to measure these soil attributes and they are highly location-specific, so that variability is great and they sometimes have little interpretive value. Moran *et al.,* (1988), however, in a study of a soil in a wet-and-dry environment, show that a soil treated with minimum tillage had more pores, identified directly by image analysis, and higher hydraulic conductivity, measured in the field, than did a similar soil traditionally cultivated. The impact of management on soil pore soil water characteristics.

Surface sealing and crusting are common in wet-and-dry climates. Sealing increases run-off and seriously reduces the amount of water infiltrating into the soil thus reducing the water held in the soil. Sealing can also increase ponding at the surface and thereby evaporation.

Infiltration rates are often reduced 1000-fold by crusting. The crust can have a skin with a conductivity of only about 0.1 mm/h, able only to accommodate the lightest rate of precipitation (fine mist) and commonly overlies a layer of poorly-aggregated material which also has a conductivity substantially lower than that of the underlying soil. Chase and Boudouresque (1989) and Chase *et al.,* (1989) illustrate the impact of crusting in the Sahel on increased run-off from soils. They show how run-off may be reduced, and the depth of wetting increased, by covering the soil surface with mulch.

These solid fractions contribute to the consistence and strength of the soil, and their packing determines bulk density. *Bulk density* is a measure of the packing or compression of the three constituents of soil. Just as the inherent bulk density of a soil will vary by 30 per cent according to its constituents, so the limiting values of bulk density for root penetration will range from about 1.4 g $cm^3$ in a soil of clay texture to 1.8 g $cm^3$ in a sandy one.

*Soil strength* is the resistance of soil to shearing or structural failure. This reflects the friction which is built up between the soil and an implement, and depends on the density, and the roughness and shape of the soil particles. The shear strength of an individual clod decreases with wetting but, more importantly, the strength of the bulk soil increases with increasing moisture to about the lower plastic limit (known to field operators as the 'sticky point'), at which each particle is surrounded by a film of water which acts as a lubricant. Soil strength drops sharply from that point to the upper plastic limit, where the soil becomes viscous. The difference between the moisture content at the upper and lower plastic limits, termed the plasticity index, is an index of the workability of the soil. A large range or high plasticity index implies a need for large amounts of energy to work the soil to a desired tilth.

# 7

# Weathering and Soils

## INTRODUCTION

Rock weathering can produce spectacular landscapes. In Monument Valley, Arizona differential rock weathering and erosion of sandstone formations results in stair-stepped topography and distinctive towers of rock. Weathering etches less resistant layers so that bedding becomes visible, and the pattern of jointing and overall rock strength varies with rock type (lithology). Talus, a blanket of material fallen from cliff walls, covers some of the lower slopes.

The thin layer of mechanically broken and chemically altered rock at the ground surface provides the substrate in which plants root, terrestrial life derives sustenance, and erosion acts to shape landscapes. This thin skin of the Earth houses the sphere of life — the biosphere. From a geomorpho-logist's perspective, erosional processes are much more effective at shaping landscapes once solid rock breaks down into transportable sediment and decomposes into unconsolidated secondary minerals or dissolved ions.

Weathering is the chemical or physical alteration of parent material (rock or sediment), and the weathering products that mantle fresh rock (or sediment) are called soil. Weathering processes influence the physical and chemical properties of weathered rock and soil, which in turn influence geomorpho-logical processes and landforms.

Soils form Earth's outer skin, the transition from its rocky interior to a gaseous atmosphere. Soils not only harbour and sustain life, they are themselves partly composed of organic matter. Plants and animals both depend on soils, and in turn influence rates and styles of soil formation. Weathering and soil formation processes create the thin layer of regolith (soil and weathered rock) that provides the foundation for terrestrial life. Physical, chemical, and biological processes all act as weathering agents pretty much everywhere, but their rates and relative importance vary dramatically among different landscapes around the world. But not all regolith was produced in situ — the glacial debris covering New England was imported by glaciers from Canada. Weathering and erosion do not necessarily progress at the same pace. Where weathering outpaces erosion, landscapes develop thick mantles of saprolite, highly decomposed rock that lost mass but not volume during weathering. Saprolite can extend hundreds of meters deep in flat-lying, slowly eroding areas of tropical Africa and South America. It can erode like loose sand yet still retain structures or features of the original parent material, like igneous dikes or sedimentary bedding planes even though its density can be half that of intact rock due to mass loss.

This chapter reviews the processes that act to weather rocks, sediment, and minerals at and near Earth's surface, and introduces the physical, chemical, and biological processes that produce soils. We will explore the dominant controls on the transformation of primary rock-forming minerals into secondary minerals and consider global and lithologic controls on weathering and the resulting influences on landforms. We will also address how soil scientists recognize diagnostic characteristics and classify soil types.

## SOIL CLASSIFICATION

Soils are classified based on field-observable chara-cteristics. Different soils and soil horizons are distinguishable by differences in one or more soil properties that reflect the degree of soil development and provide both indications of process and measures of time and landform stability. The most important soil properties observable in the field — colour, texture, and structure — reflect a soil's clay and organic matter content and its ability to hold moisture.

Soil colour indicates both mineral composition and organic matter content. Soils with high organic matter content are typically dark brown to black. Iron-rich soils are red to yellow-brown in colour in oxidizing environments and iron is blue to black or gray in reducing environments such as wetlands. A mottled pattern of red and bluish colours often indicates seasonally saturated soils that alternate between oxidizing and reducing conditions. Light colours, like white, gray and beige are generally associated with accumulations of calcite ($CaCO_3$) or leached horizons consisting mostly of residual silica ($SiO_2$). In the field, soil scientists use a book of colour charts resembling those used at paint stores to identify soil colours based on a palette of standardized colours.

Soil texture is measured in terms of the proportions of different particle sizes in the soil. The U.S. Department of Agriculture defines twelve standard soil textures based on the relative proportions of clay, silt, and sand (*e.g.*, clay loam, silt loam, silty clay, etc...). Although soils may contain larger clasts (like gravel), soil properties measured in the field generally are described based on particles less than 2 mm in diameter (*i.e.*, sand and smaller).

Soil texture greatly influences the stickiness of a soil as well as its ability to resist deformation (plasticity). A simple field test for soil composition is to moisten a sample of soil and try to make a ball in one's hand. Clay can be rolled into a "worm"; silt and sand cannot. Another simple field test — one that is not always advisable to employ — is that silt feels gritty between one's teeth, while clay has a smooth, creamy texture. Sand grains are visible to the naked eye. Soil structure describes the shapes in which soil particles cluster together. Most A horizons have granular or crumb structures in which individual clumps, or aggregates, are more or less spheroidal. In contrast, B horizons often have structures, called peds, that are shaped like blocks, prisms, or plates with rounded or angular edges. Peds can range from pea-size to fist size and reflect the deposition of clay along soil partings.

Soil organic matter includes undecomposed material like leaves, branches and bones (litter) as well as amorphous decomposed material (humus). Soil microorganisms that decompose litter into humus become active at temperatures slightly above freezing (generally >5°C), and rates of microbial decomposition increase as temperatures rise. At temperatures above about 25°C, little humus accumulates because it decomposes as fast as it is produced. Consequently, soil organic matter content tends to be greatest in mid-latitude regions with mean annual temperatures between 5 and 25°C (41 to 77°F).

## SOIL ORDERS

Soil orders are characterized by certain kinds of soil profiles. The U.S. Natural Resource Conservation Service (formerly the Soil Conservation Service) uses a system of soil taxonomy that recognizes twelve soil orders that are distinguished by diagnostic horizons and characteristics that reflect different environments, processes, and duration of formation.

Soils can be further subdivided into 64 common suborders and then into specific soil types, but the major soil orders are sufficient to characterize soil characteristics, parent material, and degree of weathering at a regional scale. Some soils are formed by factors that change over short distances, such as from differences in parent material or topography. Others reflect factors that vary over long distances (climate and vegetation types). Yet others reflect the influence of particular parent materials or time on the degree of soil development.

Immature soils and wetland soils can look the same in widely varying envrionments. Young, incompletely developed soils that are essentially raw sediment are either entisols that have a faint A horizon, or inceptisols with a weak B horizon, but not with enough horizon development to quality as one of the more developed soil orders.

Soils formed in seasonally submerged or frozen enviro-nments have distinctive characteristics. Histosols are organic-rich soils that form in wetlands where reducing conditions restrict rates of organic decomposition. Histosols are identified by their black colour, high organic matter content (more than 25 per cent), and lack of evidence for oxidation. Gelisols form in polar and subpolar environments where the ground stays frozen all year. These cold climate soils include, by definition, a layer of permanently frozen soil (permafrost) within two meters of the land surface. They typically show little evidence of weathering and have an A horizon developed over permafrost. Gelisols are often structurally disrupted and mixed by seasonal freeze/thaw processes called cryoturbation, which complicates foundation engineering in periglacial environments.

Several types of soil reflect distinctive parent materials. Andisols are formed from volcanic parent material. Vertisols are soils that contain high concentrations of swelling clay (smectite) and exhibit significant shrink/swell behaviour. Vertisols do not have distinctive horizons; they exhibit large desiccation cracks when dry, and tend to become annoyingly sticky when moist. The other six soil orders are associated with particular climate or environmental settings, and form under certain combinations of temperature and relative wetness.

## GRASSLAND SOILS

Mollisols typically develop in temperate regions with grassland vegetation, like the Great Plains of North America. Mollisols have well-developed, black, organic-rich A horizons that are often more than a meter deep and B horizons that are enriched in clay.

Grassland soils accumulate abundant organic matter because of the high seasonal production of plant litter and because grass species are generally alkaline and promote low soil acidity and therefore slow organic decomposition. Environments that permit incomplete decay of organics produce organic colloids that can leach down into the soil, making fine black A & B horizons.

This commonly happens in grasslands where the soils are too cold for decomposition in the winter, but are too dry for much decomposition in the summer. Mollisols typically contain abundant organic matter and clays with high base saturation and thus generally make fertile agricultural soils. The fertility of the grassland soils of the American Midwest is sustained by A horizons so thick and rich that they remain productive even after more than a century of agricultural erosion.

## FOREST SOILS

Surface litter produced by forests tends to be more acidic than organic matter produced by grasses or shrubs. Consequently, forest soils are generally more acidic and have A horizons that are more leached of soluble ions (notably Ca, Na, and Mg) because of extensive chelation by organic compounds. Soils that develop beneath forests are called alfisols, ultisols, and spodosols, in order of driest to wettest environment of formation. Alfisols and ultisols have thin A horizons, but the A horizon is absent in most spodosols.

Alfisols form under deciduous forests due to acidity and leaching by decaying leaves, and as such are common in the eastern United States. Alfisols are distinguished by higher base saturation and a fertile C horizon. Ultisols tend to form in old and highly weathered areas that are warm and wet and were not glaciated, reflecting more prolonged or intense weathering. Spodosols have even more extreme leaching, often with extensive E and B horizon development, mostly reflecting acidic conditions under coniferous forests due to organic acids produced by decaying needles. All three soil orders (alfisols, ultisols, and spodosols) are subject to iron, aluminum, and organic matter leaching from the A horizon and accumulating in the B horizon due to production of organic acids and prevalence of chelation.

Alfisols and ultisols form in a wide range of temperature zones, but ultisols typically characterize wetter regions and are more weathered. In ultisols, colloidal Fe and Al oxides and hydroxides may dominate as Fe and Al bearing primary minerals are nearly entirely broken down and leached. The B horizons of alfisols and ultisols are enriched in clay, whereas B horizons of spodosols are enriched in organic matter and iron and aluminum oxides. Spodosols best develop in cold, wet environments and in parent materials with high permeability, such as sands on a coastal plain. Many extensively leached forest soils have a distinctive, light coloured E horizon at the top of the B horizon that consists mostly of silica that has been leached of organic matter, $Al^{3+}$, and $Fe^{3+}$.

## ARID SOILS

Aridisols are soils that develop in dry regions. They usually have low organic matter content and an immature A horizon overlying a weak B horizon. In subtropical and arid climates, infiltrating precipitation typically evaporates within the soil profile before reaching the water table. When this occurs, ions and particles in percolating rainwater or mobilized from the A horizon precipitate or deposit in the B horizon. Calcification occurs when calcium carbonate ($CaCO_3$), gypsum, or other salts dissolved in groundwater precipitate out and coat mineral surfaces to form a distinctive white layer. The depth to this zone of evaporative accumulation reflects the amount of annual precipitation. Increasing aridity generally results in less well developed A horizons and shallower zones of carbonate accumulation.

In extremely arid settings like the Negev Desert in southern Israel, salt-rich (salic) or gypsum-rich (gypsic) horizons develop because soil water is insufficient to leach away even these very soluble minerals. Given enough time, however, red, clay-rich B horizons can develop in some of Earth's driest deserts. Dustfall is also important in the development of arid and semi-arid soils, which can have thick A horizons that develop as dust infiltrates down through a surface layer of coarse clasts known as a desert pavement.

## TROPICAL SOILS

Oxisols are the most deeply weathered of all soil orders and are characterized by highly oxidized soil horizons that form in hot, wet (tropical) environments. They generally have distinctive red colouration and little organic matter accumulation because plant and animal remains decay rapidly in such conditions. Red-coloured oxisols have an abundance of iron oxides (hematite and goethite) and tend to be depleted in exchangeable nutrient cations. The most heavily weathered tropical soils (laterites) consist of little more than residual iron and aluminum oxides such as gibbsite ($Al(OH_3)$), an aluminum oxide that forms when silica has been leached from kaolinite. The components, once heated and dried by the sun, are basically brick. Organic matter and nutrients in oxisols are in the thin O or A horizon or the vegetation, posing impressive challenges to sustaining agricultural productivity when these surficial horizons are removed by erosion.

## WEATHERING OF ROCKS

### PHYSICAL WEATHERING

Physical or mechanical processes taking place on the Earth's surface, including the actions of water, frost, temperature changes, wind and ice; cause disintegration and wearing. The products are mainly coarse soils (silts, sands and gravels). Physical weathering produces Very Coarse soils and Gravels consisting of broken rock particles, but Sands and Silts will be mainly consists of mineral grains.

### CHEMICAL WEATHERING

Chemical weathering occurs in wet and warm conditions and consists of degradation by decomposition and/or alteration. The results of chemical weathering are generally fine soils with separate mineral grains, such as Clays and Clay-Silts. The type of clay mineral depends on the parent rock and on local drainage. Some minerals, such as quartz, are resistant to the chemical weathering and remain unchanged.

### QUARTZ

A resistant and enduring mineral found in many rocks (*e.g.* granite, sandstone). It is the principal constituent of sands and silts, and the most abundant soil mineral. It occurs as equidimensional hard grains.

### HAEMATITE

A red iron (ferric) oxide: resistant to change, results from extreme weathering. It is responsible for the widespread red or pink colouration in rocks and soils. It can form a cement in rocks, or a duricrust in soils in arid climates.

### MICAS

Flaky minerals present in many igneous rocks. Some are resistant, *e.g.* muscovite; some are broken down, *e.g.* biotite.

**CLAY MINERALS**

These result mainly from the breakdown of feldspar minerals. They are very flaky and therefore have very large surface areas. They are major constituents of clay soils, although clay soil also contains silt sized particles.

**CLAY MINERALS**

Clay minerals are produced mainly from the chemical weathering and decomposition of feldspars, such as orthoclase and plagioclase, and some micas. They are small in size and very flaky in shape. The key to some of the properties of clay soils, *e.g.* plasticity, compressibility, swelling/shrinkage potential, lies in the structure of clay minerals. There are three main groups of clay minerals:

**KAOLINITES**

(Include kaolinite, dickite and nacrite) formed by the decomposition of orthoclase feldspar (*e.g.* in granite); kaolin is the principal constituent in china clay and ball clay.

**ILLITES**

(Include illite and glauconite) are the commonest clay minerals; formed by the decomposition of some micas and feldspars; predominant in marine clays and shales (*e.g.* London clay, Oxford clay).

**MONTMORILLONITES**

(Also called smectites or fullers' earth minerals) (include calcium and sodium momtmorillonites, bentonite and vermiculite) formed by the alteration of basic igneous rocks containing silicates rich in Ca and Mg; weak linkage by cations (*e.g.* $Na^+$, $Ca^{++}$) results in high swelling/shrinking potential.

**TRANSPORTATION AND DEPOSITION**

The effects of weathering and transportation largely determine the basic nature of the soil (*i.e.* the size, shape, composition and distribution of the grains). The environment into which deposition takes place, and subsequent geological events that take place there, largely determine the state of the soil, (*i.e.* density, moisture content) and the structure or fabric of the soil (*i.e.* bedding, stratification, occurrence of joints or fissures, tree roots, voids, etc.)

**TRANSPORTATION**

*Due to combinations of gravity, flowing water or air, and moving ice. In water or air:* grains become sub-rounded or rounded, grain sizes are sorted, producing poorly-graded deposits. In moving ice: grinding and crushing occur, size distribution becomes wider, deposits are well-graded, ranging from rock flour to boulders.

## DEPOSITION

In flowing water, larger particles are deposited as velocity drops, *e.g.* gravels in river terraces, sands in floodplains and estuaries, silts and clays in lakes and seas. In still water: horizontal layers of successive sediments are formed, which may change with time, even seasonally or daily.

- Deltaic and shelf deposits,often vary both horizontally and vertically.
- From glaciers, deposition varies from well-graded basal tills and boulder clays to poorly-graded deposits in moraines and outwash fans.
- In arid conditions: scree material is usually poorly-graded and lies on slopes.
- Wind-blown Löess is generally uniformly-graded and false-bedded.

## LOADING AND DRAINAGE HISTORY

The current state (*i.e.* density and consistency) of a soil will have been profoundly influenced by the history of loading and unloading since it was deposited. Changes in drainage conditions may also have occurred which may have brought about changes in water content.

## IGNEOUS ROCK WEATHERING

There is a well known expression that captures much of the story of rock weathering and illustrates the important role it has played in Earth history:

"Igneous Rocks + Acid Volatiles = Sedimentary Rocks + Salty Oceans"

What we mean by this will become clearer if we look at the details of the weathering processes. Consider a boulder or rock containing *Feldspar* minerals. Feldspar is a general term for a group of aluminosilicate minerals containing sodium, calcium, or potassium and having a lattice framework structure that makes for rigidity. Feldspars turn out to be one of the most common minerals in the Earth's crust. Feldspars are weathered through the chemical process of *hydration:*

$$K\,Al\,Si_3O_8 + H_2O \rightarrow Al_2SiO_5(OH)_4$$

In this chemical formula feldspar reacts with water to produce a kaolinite (clay). Notice that the chemical equation does not exactly balance, that is, not all the elements on the left hand side appear on the right hand side. This is because soluble elements, such as potassium (K) are *leached* out during the chemical reaction and carried away as dissolved salts. The process of leaching can perhaps best be understood by analogy with the making of coffee. When hot water is passed over crushed coffee beans, the soluble components (making the coffee) are leached away, leaving the insoluble crushed coffee bean remnants behind.

In this way, rocks containing feldspars are weakened though the conversion of rigid feldspar to more plastic clays which do not have anything like the same structural rigidity. The process occurs at exposed surfaces of the minerals making up the rock. Through geologic time, large amounts of sedimentary r0ocks have been deposited as part of this process. In fact about 75 per cent of all exposed rocks on the Earth's surface today are of sedimentary origin and have been brought to the surface by geologic uplift.

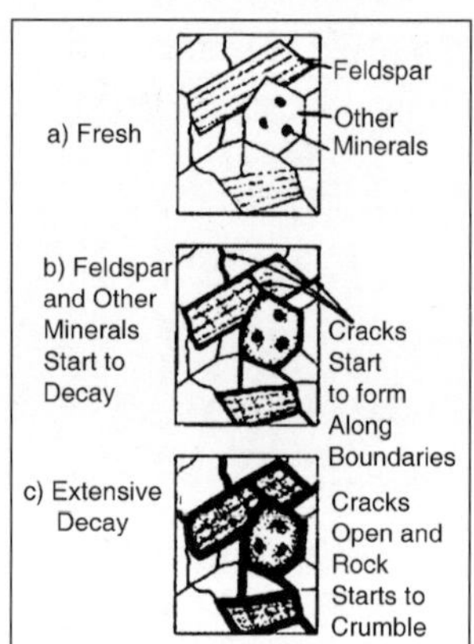


Of course, geological processes return some of the sedimentary rocks to the mantle of the Earth, where they are converted back to the primary materials under conditions of great temperature and pressure. Rock weathering is also critical for the release of biochemical elements that have no gaseous form-examples are calcium, Ca, Potassium, K, Iron, Fe, and Phosphorus, P. The latter element plays a key role in cell metabolism. Thus, we can say that weathering provides key nutrients for life through the process of leaching.

The soluble nutrients are transferred to soils layers below the immediate surface (*e.g.*, the "B" layer). This is a major reason that plants have evolved root systems-to search out these critically needed nutrients below ground. Igneous rock weathering proceeds in stages. It is useful to picture the inside of a rock-made up of interlocking minerals of irregular shapes, each being rigid. Figure shows a microscopic view of such an interior rock composition, with some of the grains being feldspars.

As weathering proceeds, the boundaries of the vulnerable feldspar (and other) mineral grains start to decay. As the decay proceeds, water can reach more and more feldspar surfaces and the process accelerates. This process can also be accelerated by melt-freeze cycles that force the grains apart due to the difference between the volume occupied by water and ice. We can see that weathering is due to the combined effects of chemical and mechanical decay.

The chemical and physical processes of weathering transform the igneous rock into sand and clay particles and dissolved salts. Chemical weathering can add carbon dioxide, water, and oxygen. The link provided courtesy of the National Park Service shows an example of a weathering rock. It is interesting to note that, since most of the Earth's exposed rock is of sedimentary origin and since sedimentary rocks are a by-product of weathering, most of the rocks that are weathering away in today's world are second or perhaps even third generation rocks.

That is, they originated as igneous material, became sedimentary through weathering and transport to the bottom of shallow waters, were then subject to geologic uplift to become again exposed and, finally, began to undergo weathering yet again. The natural world is full of such endless cycles.

## HOW FAST DOES A ROCK DECAY

The best answer to this question is... it depends. It depends on the local environment and the type of rock. For example, an iron nail buried in the ground in Michigan will only take a year or so to decay to the point that it is easily snapped in two. Iron nails rust much more slowly in drier environments. Aluminum cans decay very slowly, even in humid climates. Glass decays even more slowly, while plastic is considered essentially non-biodegradable. It is somewhat ironic that a plastic tombstone will endure much longer than one made in marble!

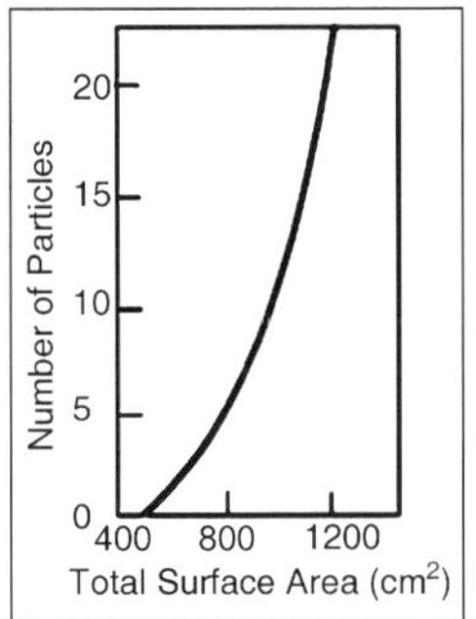


From what we have already discussed, soils themselves aid in rock decay, as do melt-freeze cycles and bacterial action. Thus, soils are a consequence of weathering, but also a factor in accelerating weathering. The production of soil is a *positive feedback* process. The following table illustrates rates of weathering for three rock types as a function of climate. As more of a rock becomes amenable to weathering, the speed of weathering increases. This can be understood if we plot the rate at which the available exposed surface area of a rock increases as the rock fragments.

## NUTRIENTS

Although living tissue is composed of carbon, hydrogen, and oxygen in the approximate proportion of $CH_2O$, as many as 23 other elements are necessary for biochemical reactions and for the growth of structural biomass.

Other important nutrients include magnesium, potassium, iron, sulphur, etc. We should note that, although carbon, nitrogen and sulphur can be obtained from the atmosphere, calcium, magnesium, potassium, iron, and phosphorus all come from rock weathering processes. The atmosphere has no store of these essential nutrients As mentioned earlier, one of the main purposes of plant root systems is to get access to nutrients stored in the soil. Some plants go to enormous lengths to do this. For example, Emiliani quotes that "A single plant of winter rye, 50 cm high, was found to have a root system consisting of 143 main roots, 35,600 secondary roots, 2.3 million tertiary roots, and 11.5 million quaternary roots! The root system was found to have a total length of 600 km and a total surface of about 250 square meters".

Delivery of nutrients into plant root systems can occur by several pathways. In some cases, direct uptake in water solution occurs. Sometimes, plants actually have to protect themselves against too much nutrient intake.

Too much of a good thing can prove poisonous. An example of this can be seen in the accumulations of calcium carbonate deposits that surround the roots of some desert shrubs.

**Table. Nature's Vitamin Requirements**

| Amount | Nutrient |
|---|---|
| 100 parts | N |
| 15 parts | P |
| 50 parts | K |
| 5 parts | Ca |
| 5 parts | Mg |
| 10 parts | S |

Some nutrients, such as nitrogen, phosphorus, and potassium are often harder for roots to find and specialized (incredibly efficient) enzymes have evolved located in root membranes to seek out these scarce and needed resources. If some nutrients are not readily available, plants will grow more slowly and/or increase their root/shoot ratio.

In general, the availability of nutrients (deficit or surplus availabil ity) often controls the form of the ecosystem, determining its overall productivity, and influencing which particular set of plants come to predominate. Excess nitrogen can lead to the loss of fine root biomass and deficiencies in other nutrients. The pool of nutrients held in the soil and vegetation is many times larger than the annual receipt of nutrients from the atmosphere and rock weathering. Thus, life *husbands* its needed nutrients on land, storing much of the total in the humus. Recycling of nutrients is critical to the productivity of natural ecosystems, although less critical to crop production, due to the availability of commercial fertilizers.

**Table.Percentage of Annual Nutrient Requirement for Growth of Hardwood Forest**

| Process (kg/ha/yr) | N | P | K | Ca | Mg | |
|---|---|---|---|---|---|---|
| Total Growth Requirement | 115 | 12 | 67 | 62 | 10 | Atmospheric |
| Inputs | 18 | 0 | 1 | 4 | 6 | Rock Weathering |
| Inputs | 0 | 13 | 11 | 34 | 37 | Reabsorptions |

| (Intra-system) | 31 | 28 | 4 | 0 | 2 | Detritus Turnover |
|---|---|---|---|---|---|---|
| 69 | 81 | 86 | 85 | 87 | | |

The data of the table came from a study of the famous Hubbard Brook ecosystem in New Hampshire. It shows how effective the soils are in storing the needed nutrients and how limited are the rates of supply from atmosphere and rock weathering.

Acid rain caused by human emissions of nitrogen and sulphur oxides leads to enhanced weathering and changes in nutrient ratios. For example, recent studies suggest that forest growth has declined in areas downwind of air pollution. Acid rain appears to increase the movement of aluminum ions, which may, in turn, reduce the intake rates of calcium and other nutrients.

In the oceans, life is also limited by the availability of nutrients. The productivity is highest on continental shelves and in regions of upwelling. Nutrients are removed from surface waters by downward sinking and are regenerated in deep waters. A paucity of nutrients limits production in the open oceans. On the other hand, marine productivity is threatened by excessive human inputs of nitrogen and phosphorus in coastal regions, where it leads to excessive algal blooms, whose subsequent decay robs deeper water of oxygen, creating "dead zones".

## BIOLOGICAL WEATHERING

Biological activity can greatly influence both physical and chemical weathering. Organisms directly affect mineral weathering by mechanically breaking down rocks and indirectly by catalyzing chemical reactions and causing changes in environmental conditions (such as changing soil pH). The growth of plant roots helps disintegrate rocks by gradually prying apart cracks, fractures, and other openings.

Burrowing animals, like gophers, ants, and termites excavate rock fragments and mix them into overlying soils. The process of biologically mediated mixing, called bioturbation, disrupts original structures or fabric in the parent material. Biological activity also increases the potential for chemical weathering by increasing surface area and creating and enlarging pathways for water flow.

Because oxygenated water is reactive, the more water that moves through a soil the more weathering happens. Decaying organic matter and respiration of soil microorganisms and plant roots indirectly influence chemical weathering by increasing the concentration of $CO_2$ in soil gases, thereby promoting acidification of soil water and enhancing chemical weathering in the fractures and pores through which soil water flows.

Plant roots release organic acids that enter soil water and either attack fresh mineral surfaces directly or exchange hydrogen ions for nutrient cations that plant roots absorb as they take up water. Decaying organic matter also releases humic acids that facilitate further weathering. Bacteria can mediate weathering reactions and lichen colonisation is the first step in weathering of many rocky surfaces.

Evapotranspiration by plants can also strongly affect the amount of soil water available for weathering, an effect that is particularly important in desert climates. In short, more life leads to more breakdown and thus more weathering.

## PHYSICAL PROCESSES OF WEATHERING

Physical processes mechanically break rocks into smaller pieces (disaggregation), and chemical processes alter their mineral composition (decomposition). During physical weathering, rocks and rock-forming minerals break apart without changes in composition. In contrast, chemical weathering involves reactions that change primary rock-forming minerals into secondary minerals, such as clays. In the process some elements are lost in solution to surface and groundwater. The two types of weathering (physical and chemical) are related. The greater surface area that results from breaking rocks into smaller fragments accelerates chemical weathering, and changes in mass or volume accompanying chemical decomposition can promote physical weathering. So even though different environmental conditions tend to favour physical versus chemical weathering, the processes are complementary and often strongly influence each other.

Biological activity catalyses both physical and chemical weathering, and depends on the active species, substrate, and weathering processes in question. Tree roots, for example, slowly pry rocks apart as they grow and the organic acids they secrete promote chemical weathering. And in many environments the soil is a nano-scale jungle of microbial activity that promotes weathering.

### PHYSICAL WEATHERING

Physical weathering reduces parent material into smaller pieces, and in some cases even fragments of individual mineral grains. Zones of weakness in the original material, like cleavage or bedding planes, metamorphic foliation, and even mineral boundaries often determine the size and shape of rock fragments. Cracks form where stresses imparted by expansion, contraction or shearing exceed the strength of rocks or minerals. The most pronounced physical weathering in bedrock occurs where there is a strong directional contrast in pressure, as is the case at and near the land surface where overburden is minimal and open fractures are more abundant than at depth where confining pressures are greater and more uniform.

Important general mechanisms of physical weathering include release of confining pressure that allows rock to expand and fracture, thermal expansion from heating by insolation or forest fires, and cyclic expansion and contraction from freeze/thaw action in cold environments. Wetting and drying is important in rocks with minerals susceptible to shrinking and swelling. And expansion by growth of salt crystals can crack rocks in arid environments.

**EXFOLIATION**

Erosion of overlying material unloads rocks, allowing them to expand upward and crack as they are exhumed. Because vertical confining stresses lessen as the weight of overlying rock decreases, rocks that are brought up from depth by tectonic uplift and unburdened by erosion tend to break into thin, rind-like sheets that are oriented perpendicular to the direction of stress release (normal to the minimum principle stress).

These exfoliation sheets are typically 1 to 10 m thick and generally follow the shape of the land surface, resulting in onion-like fracture patterns. Exfoliation sheets are most apparent on bare rock surfaces, but slope-parallel fractures also develop within soil-mantled slopes, creating networks of near-surface discontinuities that can greatly influence near-surface hydrology and slope stability.

Because exfoliation sheets are more readily eroded than underlying unfractured rock, slope-parallel fractures tend to mimic topographic forms as subsequent sheets form and erode. The process of exfoliation commonly produces dome-shaped surfaces, like Yosemite Valley's famous granite Half Dome.

Exfoliation is typically better developed in igneous and metamorphic rocks that formed deep within Earth's crust than in sedimentary and volcanic rocks formed at shallow depths. Sedimentary rocks form within a few kilometers of Earth's surface. In contrast, igneous and metamorphic rocks usually form at depths exceeding ten kilometers, and are commensurately stronger due to higher temperatures and pressures of formation.

As erosion exhumes once deeply buried rocks, bringing them closer to Earth's surface, it increases the contrast between the stresses locked in during crystallisation or metamorphism and the stress imposed by adjacent rock. When the difference becomes greater than the strength of the rock, it cracks and sets the stage for exfoliation.

Even in sedimentary rocks, upward expansion from unloading separates bedding planes, allowing outward expansion where bedding tilt matches the topographic slope. Fracturing due to stress release is an important mechanism by which water, oxygen, and plant roots penetrate into a rock mass. Rock masses subject to tectonic stresses or that have shrunk upon cooling develop patterns of joints, fractures along which no significant movement has occurred, but that provide planes of weakness along which weathering and erosion can penetrate and preferentially remove material. Because the tensile (extensional) strength of rock is typically many times less than its compressive or shear strength, jointing is often well developed even in relatively strong, erosion resistant rocks.

Parallel sets of extensional joints develop orthogonal to the direction of maximum stress as a result of either crustal extension or cooling of igneous rock. Joint systems can have strong topographic expression in arid and semi-arid landscapes where bedrock properties dominate hillslope morphology. Joints provide avenues for water and plant roots to promote weathering that focus erosion and facilitate infiltration and groundwater flow into rock that, in turn, promote more aggressive weathering.

Erosion along intersecting sets of joints can produce isolated columns of rock separated by weathered out joints. Joints also form parallel to the strike of bedding as a result of bending or folding of brittle rocks. Lithology greatly influences the degree of jointing, with joints typically being better developed in more brittle rocks such as sandstone and granite and less well developed in more flexible rocks like shale. In addition, joint patterns developed in igneous rocks tend to be less linear and more irregularly spaced than those developed in sedimentary rocks.

## THERMAL EXPANSION

Rocks and minerals expand when subjected to heat, but low thermal conductivity generally prevents the zone of high heating from penetrating more than a few centimeters into rocks over the course of a forest fire or a day of exposure to intense sunlight. Low thermal conductivity explains why rock surfaces retain heat at night better than do vegetated surfaces and why stone buildings stay relatively cool in summer and warm in winter without the need for additional insulation. Extreme temperature differences between a rock's surface and its cool interior can produce large differential stresses that result in spalling of a thin outer layer that can be up to several centimeters thick. The role of daily temperature fluctuations due to heating by sunlight has long been debated, but recent work has documented convincing evidence that daily heating and cooling can lead to rock fracture. For example, it has been shown that the majority of cracks in desert rocks that are not related to rock heterogeneities like bedding are oriented north-south.

That the majority of vertical cracks in rocks in deserts around the world are preferentially oriented north-south suggests that thermal stresses produce north-south oriented cracks due to differential heating when the sun crosses the sky from east to west. The coefficient of thermal expansion of certain minerals such as calcite can cause a rock to break apart along mineral boundaries as individual grains expand and contract with small changes in temperature. The weathering product of this type of granular disintegration, known as grus, is essentially loose sand made of individual mineral grains.

## FREEZE-THAW

The expansion of water in fractures and pore spaces as it freezes into ice is particularly effective at breaking rocks apart. Consequently, frost shattering is a primary weathering process in alpine and polar environments that are subject to frequent freeze-thaw cycles. Water expands by almost ten per cent when it freezes, so rock generally will not shatter unless about 80 per cent of the available pore space is saturated. Ice simply expands into partially saturated voids without generating high pressures on the surrounding rock.

However, hydrofracturing where freezing of water proceeds from the outside in forces water into the tiny ends of fractures, producing an effect like a hydraulic jack. In addition, as ice lenses form, water and vapor flow towards them, imparting enough force that growing ice crystals crack rocks. Many alpine slopes in environments subject to frost shattering are covered by a blanket of rock blocks called felsenmeer (German for sea of rocks). The rubble-covered summit of Mt. Washington in New England is a great example.

## WETTING AND DRYING

The addition and removal of water from minerals, processes known as hydration and dehydration, can cause swelling or contraction capable of fracturing rocks or disaggregating them into individual mineral grains. The volume increase in the anhydrite to gypsum transformation, in particular, has been ruinous to archaeological monuments. Most micas and clay minerals exhibit some shrink when they dry and swell when they absorb water.

Some can expand to twice their original volume, and certain particularly susceptible clays, like smectite (montmorillonite), swell several-fold when they get wet. Expansive soils that are composed of minerals with shrink/swell behaviour cause substantial engineering problems and extensive damage through cracked foundations annually in the United States, primarily in the South and the western plains wherever such soils are common. Swelling of bentonite-rich soils in Montana, Wyoming, and Colorado can make travel on dirt roads nearly impossible after rainstorms as sticky mud bogs down even 4-wheel drive vehicles.

In some rocks, like granite, expansion of biotite micas during weathering to hydratable vermiculite clay can pry apart grain boundaries, and produces grus. Similarly, cycles of wetting and drying can lead to repeated expansion and contraction that disaggregates micaceous sandstone and turns hard, erosion-resistant rock into a loose pile of sand in a single season. In rock types susceptible to this process, exposure to seasonal cycles of wetting and drying leads to rapid rates of bedrock erosion through the annual formation of a loose crust or outer covering of highly erodible material easily removed by subsequent high flows.

Hydration and expansion of salts such as gypsum or halite within pore spaces also cause spalling and rock disintegration. Salts can expand several fold when hydrated. Repeated wetting and drying, as well as the growth of salt crystals due to evaporation of fluids in near-surface fractures and pore spaces, can gradually pry rocks apart in arid climates. Salt weathering may even be an important mechanism of rock disintegration in the dry desert landscapes on Mars.

## CHEMICAL WEATHERING AND SOIL

Chemical weathering occurs because the minerals in rocks form at deep Earth pressure and temperature conditions that are not in equilibrium with conditions at Earth's surface, and are thus vulnerable to chemical decomposition and transformation. A primary weathering agent is rainwater that percolates into the ground and promotes chemical weathering because it contains dissolved ions gained from the atmosphere and from the soils through which it moves. The bipolar water molecule is a potent solvent — given time. The metabolism of soil microorganisms and decay of organic matter enhance weathering as they add organic acids to water moving through soils.

Root respiration and microbial oxidation make a soil atmosphere rich in $CO_2$. The addition of water makes carbonic acid. Sulfuric and nitric-acid weathering are important in some areas and always present at some level. In addition, biochemical activity tends to increase rates of chemical reactions between soil fluids and minerals by lowering pH and increasing temperature.

Processes of chemical weathering include solution, oxidation and reduction, hydrolysis, ion exchange, and the formation of new, more stable secondary minerals, like clays and hydrous oxides. The end results of chemical weathering depend on a variety of interacting factors including the composition and texture of the parent material and the chemical, physical, and biochemical processes acting in a particular environment.

The mobility and stability of the secondary minerals and solutions produced depend on environmental conditions like pH, Eh, and temperature. Chemical weathering is the breaking of chemical bonds — metallic, ionic, and covalent. The corresponding principle weathering processes are electron exchange (oxidation/reduction), ionisation (solution), and ion exchange (as in acid attack).

As many rocks are dominated by a mix of ionic and covalent bonds, solution and acid attack are major weathering processes. Hydration and dehydration are also important weathering mechanisms for certain rock types and in certain environments. Chemical weathering is essential for the biosphere in the critical zone, where vegetation demand for Ca, Mg, K, $NO_3$, and P is extraordinary. Nutrients derived from minerals are cycled through ecologicai systems because of the slow pace of weathering or the depleted nature of surficial materials (especially in key limiting elements such as phosphorus).

### OXIDATION/REDUCTION

Oxidation is a process during which an element loses an electron to a receptor, often an oxygen ion, like when iron rusts. Conversely, reduction is defined as the gain of an electron. Free oxygen is rare at crustal depths where rocks form, but abundant at Earth's surface. Rocks and rock-forming minerals typically oxidize when they are exposed to well-oxygenated soil water, directly to the atmosphere, or to gases in soil pores.

Reducing conditions in oxygen-poor waters with lots of organic matter, like swamps and peat bogs with high seasonal water tables, generally prevent oxidation, retard organic decay, and slow down weathering. When soils alternate between saturated and unsaturated conditions, a speckled colour pattern known as mottling develops — with gray colours due to the lack of oxidized iron as well as reddish colours due to oxidation.

Oxidizing potential is expressed in terms of redox potential (Eh), the availability of free oxygen, which is greatly influenced by the amount of dissolved organic matter in pore fluids. Soils typically have Eh high enough to oxidize most common elements, but iron, manganese and sulfur are especially prone to rapid oxidation and typically occur as red, black, and yellow coatings in soils.

Redox potential exerts a substantial influence on ion mobility, and oxidation is often the first form of weathering to alter freshly exposed rock surfaces. Oxidized versions of elements are relatively immobile, whereas reduced versions are far more mobile. Over time, rinds of oxidized material form on the surfaces of outcrops or boulders as other material is leached away.

When rocks containing common iron bearing carbonates, sulfides, and silicates (such as olivine and biotite) oxidize, they become susceptible to additional physical weathering. Oxidation produces relatively insoluble ferric oxides, like hematite ($Fe_2O_3$), or rust, and oxyhdroxides like goethite (FeO(OH)) that colour soils and weathered rock various shades of reddish or yellowish brown. By the same token, oxygen starved conditions (such as those under stagnant water rich in decomposing organic matter) turns iron and manganese back into reduced forms, which allows them to be dissolved and leached if the water drains or is flushed from the soil.

## SOLUTION

The flow rate and acidity of pore water are two of the most important factors influencing the amount of disolution from soil, sediment, or rock in a given weathering environment. In addition, pH strongly affects the solubility of most elements. Rainwater is slightly acidic from dissolved atmospheric $CO_2$, and chemical and biologic weathering processes often act to lower the pH of water moving through a weathering zone. In weathering zones with active groundwater circulation, fresh water comes in contact with parent material, and weathering continues as leaching removes dissolved material. Slowly circulating pore waters retard dissolution as the number of dissolved ions in solution approaches an equilibrium concentration.

Natural groundwater tends to be slightly acidic due to dissolution of carbon dioxide in water to produce carbonic acid ($H_2CO_3^-$). Carbonic acid is not a strong acid but it is extremely abundant because it forms wherever water encounters $CO_2$ through the carbonation reaction:

$$H_2O + CO_2 \text{ « } 2H_2CO_3^-$$

Decay of organic matter together with respiration of soil invertebrates, bacteria, and root systems can elevate $CO_2$ concentrations in soil pores so that they are 10 to 100 times greater than atmospheric concentrations. This makes carbonation a particularly important factor in heavily vegetated areas. Cold temperatures also favour formation of carbonic acid in soil water, because the solubility of $CO_2$ is inversely proportional to temperature, as is true of most gases.

*In solution, carbonic acid readily disassociates into hydrogen ($H^+$) and bicarbonate ions ($HCO_3^-$):*

$$2H_2CO_3^- \leftrightarrow H^+ + HCO_3^-$$

Consequently, bicarbonate is the most common cation in natural groundwater. Atoms exposed on the mineral surfaces of rock and soil particles are electrically charged ions that react with dissociated hydrogen ($H^+$) and hydroxide ($OH^-$) ions in water.

This interaction breaks bonds, effectively disassociating individual mineral molecules and causing exchanges that release cations from the mineral surface into solution. Mineral structures become unstable and vulnerable to further weathering when they lose cations, so a little weathering promotes more weathering. Congruent dissolution occurs when all the constituents of an individual molecule are separated and remain in solution.

During incongruent dissolution, some of the released ions recombine to create new compounds and secondary minerals. Dissolved material may remain in solution and move along with flowing water and re-precipitate elsewhere, or it may emerge into streams and rivers, and eventually reach the ocean. The origin of the salinity of the oceans lies in the long-term delivery of dissolved material in stream water. Most common elements are soluble to some degree in both rainwater and soil water. Consequently, water circulation promotes solution by introducing fresh water that removes dissolved ions from mineral surfaces.

The dissolution of calcite ($CaCO_3$, calcium carbonate) is a particularly important chemical weathering reaction. This occurs in the presence of carbon dioxide ($CO_2$) and introduces bicarbonate ions ($HCO_3^-$) into solution. The resulting reaction is expressed as,

$$CaCO_3 + H_2O + CO_2 \leftrightarrow Ca^{++} + 2HCO_3^-$$

The carbonate dissolution reaction is reversible. An increase in $CO_2$ concentration within soil gasses, a decrease in pH, or dilution will drive the reaction to the right (as written above); the rate of carbonate dissolution will increase and the bicarbonate concentration in groundwater will go up. This effect helps percolating water erode fractures and form extensive cave systems typical of regions underlain by carbonate rocks (limestone or dolomite). Conversely, decreased $CO_2$ concentration, increased pH, or evaporation will drive the reaction to the left and favour precipitation of calcium carbonate ($CaCO_3$). It is this reaction that deposits stalagmites and stalactites in caves, as well as calcite in desert soils. When carbonic acid dissociates to form an "acid" of protons, the resulting weathering of aluminosilicate minerals consumes $CO_2$ and thus helps to cool global climate through the general reaction

$$\text{aluminosilicate} + H_2O + CO_2 \rightarrow \text{clay mineral} + \text{cations} + OH^- + HCO_3^- + H_4SiO_4$$

Earth's long-term climate is thus mediated by organic matter burial and carbonate formation (both of which sequester carbon in the geologic record) and silicate weathering, which consumes $CO_2$ (producing bicarbonate).

Over the long run, glaciations and the anthropogenic contribution to atmospheric $CO_2$ are short-term perturbations of this geologic control on global climate through carbonate burial and the influence of weathering on the concentration of $CO_2$ in atmosphere. Calcite and salts are readily dissolved in water, so carbonate rocks and evaporites are particularly susceptible to dissolution, especially in regions with abundant precipitation.

In contrast, quartz and most other rock-forming silicate minerals are not very soluble at typical Earth surface conditions, leading to slow rates of dissolution in most environments. Solubility varies greatly between minerals, but even the least soluble minerals will dissolve over time if exposed to solutions refreshed by water circulation.

Iron and aluminum oxides, however, are virtually insoluble under oxygenated soil water conditions, so these compounds are typically left behind as more soluble, mobile material is depleted. Consequently, the abundance of iron and aluminum oxides increases as rocks and sediment are exposed to weathering.

The bright red soils of the southeastern United States are an example of how these oxides can accumulate through time. Dissolution can play an important role in increasing pore space and thus increasing the percolation of water, soil acids, oxygen, and bacteria into the regolith.

## MINERAL STABILITY

The general susceptibility of rock minerals to weathering is the inverse of the sequence in which they form deep within the earth. Rocks that formed at the greatest temperatures and pressures are furthest from equilibrium at surface conditions and are therefore most susceptible to weathering when exposed to the elements. Among the common silicate minerals that account for the majority of rock forming minerals, olivine and pyroxene are most susceptible to weathering, followed in order of decreasing susceptibility to breakdown by amphibole, biotite, feldspar, muscovite and quartz. This progression, known as Goldich's Weathering Series, is the opposite of the order of crystallisation as magma cools, familiar to geologists as Bowen's Reaction Series.

Under similar environmental conditions, rocks composed of more mafic iron- and magnesium-rich minerals (olivine, pyroxene, amphibole, and biotite) will weather faster than those composed of more felsic minerals (feldspar, muscovite, and quartz).

However, it is not always as simple as this. Due to the importance of both covalent and ionic bonds, silicates with complicated mineral structures break down more readily than do those with simpler structures like quartz ($SiO_2$) or zircon ($ZrSiO_4$), a very stable mineral even though it has a very high melting temperature. So both complexity and formation conditions are central to mineral stability.

The mobility of cations in rock-forming minerals varies greatly, and influences the relative ease and order in which weathering strips cations from rocks and secondary minerals, with the sequence from most to least mobile proceeding as $Ca^{2+}$, $Na^+$, $Mg^{2+} > K^+ > Fe^{2+} > Si^{3+} > Fe^{3+} > Al^{3+}$. The most mobile cations ($Ca^{2+}$, $Na^+$, $Mg^{2+}$) are readily stripped from mineral surfaces, tend to remain in solution, and are the first to be lost from rocks as they weather. The least mobile cations ($Si^{3+}$, $Fe^{3+}$, and $Al^{3+}$) are relatively insoluble and become concentrated in residual soils over time as weathering strips away more mobile elements.

## CHEMICAL REACTION OF HYDROLYSIS

Hydrolysis is a chemical reaction in which water molecules ($H_2O$) are split into protons ($H^+$) and hydroxide anions ($OH^-$) that react with primary rock-forming minerals to form new compounds (secondary minerals). Hydrolysis is an important chemcial weathering process that acts to break rocks apart and transform silicate minerals into weathering products.

In hydrolysis reactions, mineral cations are released into solution and replaced by hydrogen ($H^+$), producing a new mineral. This process results in the transformation of aluminosilicate minerals, like feldspars and micas, into various clay minerals. *The conversion of potassium feldspar (orthoclase) into illite or kaolinite clay are examples of hydrolysis:*

Orthoclase (feldspar) illite (clay)

$$3KAlSi_3O_8 + 2H^+ + 12H_2O \rightarrow 2K^+ + KAl_3Si_3O_{10}(OH)_2 + 6H_4SiO_4$$

Orthoclase (feldspar) kaolinite (clay)

$$2KAlSi_3O_8 + 2H^+ + 9H_2O \rightarrow 2K^+ + H_4Al_2Si_2O_9 + 4H_4SiO_4$$

Hydrolysis is not reversible. Once secondary minerals are formed, further weathering can strip additional cations and can convert secondary aluminosilicates like illite into other, more cation-depleted clays.

Upon more intensive weathering, each step in the weathering of clay minerals strips additional cations from the mineral structures. Eventually, intensive weathering can leave kaolinite, which consists of just hydrogen, aluminum, silica, and oxygen, and has no additional cations left to exchange. Progressive alteration of silicate minerals due to weathering reduces the complexity of mineral structures.

## PROCESS OF HYDRATION

Hydration describes the process in which silicate minerals combine with water or hydroxide ions ($OH^-$) to form hydrated compounds. Hydration is another way that primary minerals are converted to secondary minerals.

Common forms of hydration reactions include the conversion of anhydrite ($CaSO_4$) to gypsum ($CaSO_4 \bullet 2H_2O$), and the formation of relatively insoluble iron and aluminum hydrous oxides, like limonite ($FeO(OH) \bullet nH_2O$) and gibbsite ($AlOH_3$) in regions of intense tropical weathering like the Amazon basin.

## CLAY FORMATION

Clay minerals are both a product and a player in processes of hydrolysis and hydration. Unlike most primary minerals, with the exception of quartz, secondary minerals like clays and hydrous oxides are chemically stable under earth surface conditions. They become a major constituent in soils because their relative stability and immobility leaves them as common in situ byproducts of weathering.

Most clay minerals are layer silicates composed of sheets of alumina octahedra (an atom of aluminum bonded to six atoms of oxygen) or silica tetrahedra (an atom of silica bonded to four atoms of oxygen). Both tetrahedral (T) and octahedral (O) layers are organized around central Al, Fe, or Mg cations.

These sheets generally are bonded together in either a 1:1 structure (TO) in which each layer of alumina octahedra is paired with a layer of silica tetrahedra, or in a 2:1 structure (TOT) in which each octahedral layer is sandwiched between two tetrahedral layers. These building blocks are themselves interlayered and bound together by shared ions between the sheets. Layer architecture (1:1 vs. 2:1) and the ions between the sheets determine the physical properties of different clay minerals.

Adjacent layers in kaolinite, a clay mineral with a 1:1 structure, are held together by ionic bonds that are strong enough to prevent cations or water from entering the spaces between the sheets. Because it has few exchangeable cations held between its layers, kaolinite does not swell much when wetted, and has low plasticity (and thus little capacity to be molded).

Clays with 2:1 layer structures exhibit much more variability in the chemical composition of their octahedral sheets (typically due to substitution of $Fe^{2+}$ and $Mg^{2+}$ for $Al^{3+}$) and in the abundance and type of ions present between layers. Smectite clays, like montmorillonite, have weak bonds between the silicate layers, which allows water and ions to readily penetrate the crystal structure. Also known as swelling clays, smectites expand readily upon wetting and are a main component of expansive soils.

In expanding clays there are layers of water in the interlayer position, which explains why they can so readily take on or lose water and why they are so weak when expanded. Illite, the most common clay mineral in soils, has a strongly bonded 2:1 structure and fewer exchangeable cations between its layers, so it has less swelling potential than smectite.

Weathering of secondary minerals involves stripping off layers of silicate structure. The modification of muscovite (mica) to illite (clay), both of which consist of TOT "sandwiches" involves removal of the interlayer cations. Extreme weathering conditions can go beyond leaching of the intermediary cations and remove one of the two T layers, leaving a TO sequence of silicate layers, which is kaolinite clay.

Stripping the remaining tetrahedral layer leaves the basic octahedral layer of gibbsite. In general, smectites weather to illites, and ultimately become kaolinite. Deeply weathered soils generally have high concentrations of kaolinite. It is worth noting, however, that kaolinite can form directly from primary minerals, depending on the parent material, climate, and intensity of weathering.

## PROCESS OF CHELATION

Chelation is the process through which relatively immobile metal ions, like iron and aluminum, are rendered mobile by soluble organic compounds that form ring structures around metal ions, making them susceptible to solution and transport. Chelation is particularly important in moving iron and aluminum, which are otherwise immobile in most soils. Chelation is facilitated by organic acids (particularly fulvic acid) produced by the breakdown of soil organic matter and by lichens, which produce chelating agents that accelerate weathering and liberate nutrients that help sustain the lichen.

Iron and aluminum mobilized by chelating agents may be carried along in solution with soilwater flow until concentration changes or microbial actions break the chelating agent, causing the metal to reprecipitate. Incomplete conifer needle decay in cool, moist environments is a common source of chelating agents, leading to the stripping of Fe and other elements from forest soils.

## CATION EXCHANGE

An important outcome of chemical weathering is the ability of secondary minerals to exchange cations with soil water, thereby making nutrients available to plants. Clay minerals loosely hold exchangeable cations adsorbed on their surfaces. In many soils, the exchangable cations are associated mainly with organic matter. Ion exchange is the process by which ions in solution substitute for ions on mineral surfaces. Exchangeable cations are readily taken up in soil fluids, and they provide the dominant source of mineral nutrients for plants.

Clays and organic compounds vary in their ability to adsorb and release cations, a property called cation exchange capacity. Ion exchange is controlled by cation exchange capacity as well as by the ionic composition and pH of soil water. Strongly acidic (low pH) pore fluids allow $H+$ to substitute for and replace metal cations. As hydrogen ions exchange places with nutrient cations held on a clay surface, the number of potentially exchangeable cations decreases.

The degree to which the exchange sites are occupied by exchangeable cations other than $H^+$ and $Al^{3+}$ is called base saturation. Cation exchange progressively lowers base saturation in clay minerals by removing cations from between clay sheets. A clay with a high cation exchange capacity but a low base saturation has had its balancing cations stripped out and replaced by hydrogen ions through substantial chemical weathering.

Such clays are typically found in tropical regions with high temperature and rainfall, such as Hawaii. The progressive loss of exchangeable cations reduces soil fertility — older, more intensively weathered soils are less fertile.

## GLOBAL PATTERNS OF WEATHERING

The presence, amount, and phase of water (*i.e.*, vapor, liquid, or ice) strongly influence both chemical and physical weathering. It is not surprising, then, that global patterns in the style and intensity of weathering generally track regional climate.

Differences in mean annual temperature and precipitation control the magnitude and relative importance of physical and chemical weathering processes. Chemical weathering is strongest in regions with high year-round temperatures and abundant precipitation, namely the equatorial tropics, and weakest in cold, dry areas, like the poles.

Physical weathering is least active in dry environments and most active in regions where temperatures drop below freezing for part of the year. Consequently, tropical regions are generally dominated by chemical weathering and high latitudes by physical weathering.

Similarly, topography also greatly influences weathering, as high mountains promote mechanical weathering and low-relief environments favour the dominance of chemical weathering. Rates of physical and chemical weathering, however, are correlated because high rates of mechanical breakdown promote higher rates of chemical decomposition by exposing greater surface area to chemical attack and vice-versa (through reduced material strength).

Consequently, the highest rates of rock weathering typically occur in environments conducive to both physical and chemical weathering.

Styles and rates of weathering vary among rock types because of differences in chemical (mineralogical) characteristics and physical attributes (strength and fracturing). In particular, the ease with which water is able to penetrate rocks strongly influences weathering rates.

Mineral composition is a primary factor in chemical weathering. Some rock types, like limestone and marble, are particularly vulnerable to chemical weathering in humid climates due to the fact that they can be dissolved by weakly acidic soil water or rainwater.

But these rock types can be quite erosion resistant in arid regions where water is not sufficiently abundant to carry away dissolved ions. In contrast, quartz is far more resistant to both chemical and physical weathering, and rocks like quartz sandstone (quartz arenite) and quartzite weather more slowly than most other rocks regardless of environmental conditions.

# 8

# Financing Wastewater Management

Urban sanitation is a priority issue for cities everywhere. Major deficiencies in the provision of this basic service contribute to environmental health problems and the degradation of scarce water resources. The rapid growth of cities and the accompanying concentration of population leads to increasing amounts of human wastes that need to be managed safely. The relative success in providing cities with usable water has led to greater volumes of wastewater requiring management, both domestic and industrial. As population densities in cities increase, the volumes of wastewater generated per household exceed the infiltration capacity of local soils and require greater drainage capacity and the introduction of sewer systems.

Wastewaters flowing out of cities can, in turn, affect downstream water resources and threaten their sustainable use. The mix of problems and the capacity to deal with these sanitation problems varies amongst cities and countries. Confronting these problems requires an ability to face a number of challenges, including different environmental health challenges as well as financial, institutional and technical challenges.

## THE CHALLENGES OF URBAN SANITATION

**Table. Economic - Environmental Typology of Urban Sanitation Problems**

| Urban sanitation problems | Lower-income countries (< US$ 650 per capita) | Lower middle-income countries (US$ 650-2,500 per capita) | Upper middle-income countries (US$ 2,500-6,500 per capita) | Upper-income countries (> US$ 6,500 per capita) |
|---|---|---|---|---|
| Access to basic sanitation services | Low coverage, especially for urban poor; mainly non-sewered options | Low access for urban poor; increasing use of sewerage | Generally acceptable coverage; higher sewerage levels | Good coverage; mainly sewered |
| Wastewater treatment | Virtually no treatment | Few treatment facilities; poorly operated | Increasing treatment capacity; operational deficiencies | Generally high treatment levels; major investments over past 30 years |
| Water pollution issues | Health problems from inadequate sanitation and raw domestic sewage "ir the streets" | Severe health problems from untreated municipal discharge | Severe pollution problems from poorly treated municipal and mixed industrial discharges | Primarily concerned with amenity value and toxic substances |

The environmental health challenges facing the urban sanitation subsector in developing countries are of two types. First, there is the "old agenda" of providing all urban households with adequate sanitation services. Second, there is the "new agenda" of managing urban wastewater safely and protecting the quality of vital water resources for present and future populations. The relative importance of each agenda normally depends upon the level of development, although these two "agendas" coexist in most cities of the developing world, even in some of the most modern cities.

## BASIC SANITATION SERVICES FOR URBAN HOUSEHOLDS

The provision of sanitation services, including sewerage, has not kept pace with population growth in urban areas. Despite this, the significant progress that was achieved by countries during the 1980s has resulted in a 50 per cent increase in the number of urban people with adequate sanitation facilities. These achievements, although impressive, were not sufficient because the number of people without adequate sanitation actually increased by 70 million in the same period, and as many remained unserved as were provided with service. The health consequences of the service shortfalls are enormous and fall most heavily on the urban poor. In most low-income communities, the pollutant of primary concern is human excreta.

It has been reported by WHO that 3.2 million children under the age of five die each year in the developing world from diarrhoeal diseases, largely as a result of poor sanitation, contaminated drinking water and associated problems of food hygiene. Infectious and parasitic diseases linked to contaminated water are the third leading cause of productive years lost to morbidity and mortality in the developing world. Diarrhoeal death rates are typically about 60 per cent lower among children living in households with adequate water and sanitation facilities than those in households without such facilities.

Looking to the future, the challenge of the next two decades dwarfs the progress made in the past decade; some 1,300 million new urban residents will require sanitation services in addition to those presently without service. In total, this is roughly six times the increase in service provided during the 1980s. Clearly, the aim of providing all urban households with adequate sanitation services still poses large financial, institutional and technical challenges.

## URBAN WASTEWATER MANAGEMENT AND POLLUTION CONTROL

A "new agenda" of environmentally sustainable development has emerged forcefully, and appropriately, in recent years. One aspect of sustainable development is the quality of the water environment which is seen as a global concern about sustainable water resources. The situation in cities in developing countries is especially acute. Even in middle-income countries, sewage is rarely treated. Buenos Aires, for example, treats only 2 per cent of its sewage, a percentage that is typical for the middle-income countries of Latin America. There is also the problem of uncontrolled industrial discharges into municipal sewers, increasing organic loads and introducing a range of chemical contaminants that can damage sewers, interrupt treatment processes, and create toxic and other hazards.

Water quality is far worse in developing countries than in industrialised countries. Furthermore, while environmental quality in industrialised countries improved through the 1980s, it did not improve in middle-income countries, and even declined sharply in lower-income countries. The costs of this degradation can be seen in many ways. The vast majority of rivers in and around cities in developing countries are little more than open sewers. Not only do these degrade the aesthetic quality of life in the city, but they constitute a reservoir for cholera and other water-related diseases. The cause of the major outbreak of cholera in Peru in 1991 could be traced to inadequate urban sanitation and water contamination. It cost the Peruvian economy over US$ 150 million in 1991-92 in direct and indirect health impacts. Similarly, the otherwise inexplicable persistence of typhoid in Santiago over four decades has been attributed to the pollution of irrigation waters by untreated metropolitan discharges.

Energetic emergency measures, taken as a result of the Latin American cholera outbreak in 1991, prevented the spread of cholera in Santiago and brought typhoid under control with estimated savings in direct and indirect health costs in the order of US$ 77 million. The costs of urban water pollution also create an additional burden for cities in the form of higher water supply costs. In metropolitan Lima, for example, the cost of upstream pollution has increased water treatment costs by about 30 per cent. In Shanghai, China, water intakes had to be moved upstream more than 40 km at a cost of about US$ 300 million.

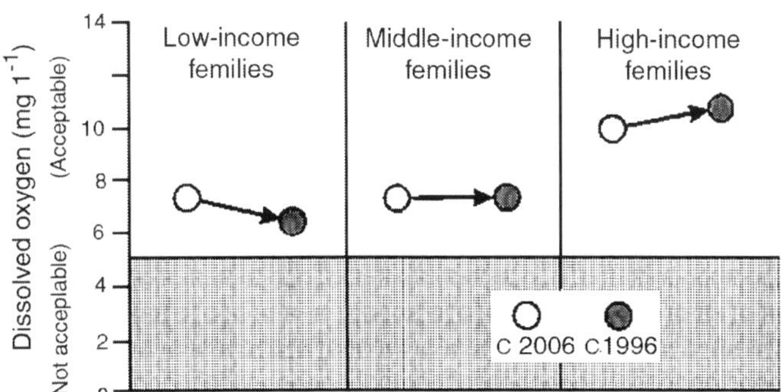


**Fig.** Dissolved Oxygen Concentrations in Rivers in Developing and Developed Countries.

## CONNECTION BETWEEN SANITATION SERVICES AND ENVIRONMENTAL ISSUES

To understand the connection between sanitation services and environmental issues, it is necessary to consider the sequence in which people demand water supply and sanitation services. For a family which migrates into a shanty-town, the first environmental priority is to secure an adequate water supply at reasonable cost. This is followed shortly by the need to secure a private, convenient and sanitary place for defecation. Families show a high willingness to pay for these household or private services, in part because the alternatives are so costly. Accordingly, they pressure local and national governments to provide such services, and in the early stages of economic development much external assistance goes to meeting the strong demand for these services.

The very success in meeting these primary needs, however, gives rise to a second generation of demands, namely for the removal of wastewater from the household, then from the neighbourhood and then from the city. As cities succeed in meeting this demand another problem arises, namely the protection of the environment from the degrading effects of such large and concentrated pollution loads. Thus it is no surprise that the portfolio of external assistance agencies has focused heavily on the provision of water supply.

For example, World Bank lending for water and sanitation over the past 30 years has only included about 15 per cent for sanitation and sewerage, with most of this spent on sewage collection and only a small fraction spent on treatment. In a description of the Orangi Pilot Project in Karachi, Pakistan, Hasan describes how forcefully poor people demand environmental services, once the primary demand for water supply is met, and how it is possible to respond to the challenge of these new demands.

## THE FINANCIAL CHALLENGES

Completing the supply of basic sanitation services and making progress on wastewater management and pollution control creates major financial challenges for developing countries. Mobilising the necessary financial resources requires both recognising the need for an urban sanitation subsector and reliance on new ways of financing urban sanitation, sewerage and wastewater management.

## RESPONDING TO THE DEMANDS OF HOUSEHOLDS AND COMMUNITIES

In recent years there has been a remarkable consensus on market-friendly and environment-friendly policies for managing water resources and for delivering water and sanitation services on an efficient, equitable and sustainable basis. At the heart of this consensus are three closely related guiding principles expressed at the 1992 Dublin International Conference on Water and the Environment, namely:

- *The Ecosystem Principle:* Planners and policy makers at all levels should take a holistic approach linking social and economic management with protection of natural systems.

- *The Institutional Principle:* Water development and management should be based on a participatory approach, involving user, planners and policy makers at all levels, with decisions taken at the lowest appropriate level.
- *The Instrument Principle.* Water has an economic value in all its competing uses and should be recognised as an economic good.

The challenge facing the urban sanitation subsector is to put these general principles into operation and to translate them into practice on the ground. The new consensus gives prime importance to a central principle of public finance, *i.e.* that efficiency and equity both require that private resources should be used for financing private goods and that public resources should be used only for financing public goods. Implicit in this principle is a belief that social units themselves, whether households, commercial organisations, urban communities or river basin associations, are in the best position to weigh the costs and benefits of different levels of investment.

The vital issue in the application of this principle to the urban sanitation subsector is the definition of the decision unit and the definition of what is internal (private) and external (public) to that unit. For each level, the demand for sanitation services must be understood, and each social unit should pay for the direct service benefits it receives. To illustrate the application of this emerging ideal, it is necessary to consider how urban sanitation should be financed.

## SANITATION, SEWERAGE AND WASTEWATER MANAGEMENT

The benefits from improved sanitation, and therefore the appropriate financing arrangements, are complex. At the lowest level, households place high value on sanitation services that provide them with a private, convenient and odour-free facility which removes excreta and wastewater from the property or confines it appropriately on-site. However, there are clearly benefits which accrue at a more aggregate level and are, therefore, "externalities" from the point of view of the household. Willingness-to-pay studies have shown consistently that households are willing to pay for the first category of service benefits, but have little or no interest in paying for external (environmental) benefits that they consider beyond their concern.

At the next level (*i.e.* the block) households in a particular block value services which remove excreta from the block as a whole. Moving up a level, to that of the neighbourhood, residents value services which remove excreta and wastewater from the neighbourhood, or which render these wastes innocuous through treatment. Similarly, at the level of the city, the removal and/or treatment of wastes from the city and its surroundings are valued. Cities, however, do not exist in isolation - wastes discharged from one city pollute the water supply of downstream cities and of other users.

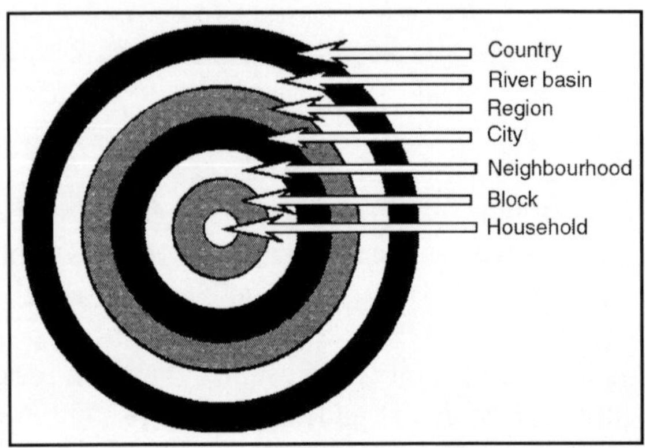


**Fig.** Levels of Dicision-making on Water and Sanitation.

Accordingly, groups of cities (as well as farms and industries and others) in a river basin can perceive the collective benefit of environmental improvement. Finally, because the health and well-being of a nation as a whole may be affected by environmental degradation in one particular river basin, there are sometimes additional national economic, health and environmental benefits from wastewater management in that basin. The example of typhoid in Santiago illustrates the latter point. The fundamental principle of public finance is that costs should be assigned to different levels in this hierarchy according to the benefits accruing at the different levels.

This suggests that the financing of sanitation, sewerage and wastewater treatment should be allocated approximately as follows:

- Households pay the cost incurred in providing on-site facilities (bathrooms, toilets, sewerage connections).
- The residents of a block collectively pay the additional cost incurred in collecting the wastes from individual homes and transporting these to the boundary of the block.
- The residents of a neighbourhood collectively pay the additional cost incurred in collecting the wastes from blocks and transporting these to the boundary of the neighbourhood (or of treating the neighbourhood wastes).
- The residents of a city collectively pay the additional cost incurred in collecting the wastes from blocks and transporting these to the boundary of the city (or of treating the city wastes).
- The stakeholders in a river basin (cities, farmers, industries and environmentalists) collectively assess the value of different levels of water quality within a basin and decide on the level of quality they wish to pay for, and on the distribution of responsibility for paying for the necessary treatment and water quality management activities.

- The nation, for the achievement of broader public health or environmental benefits, may decide to pay collectively for meeting more stringent treatment standards.

**Sanitation and Sewerage**

Although there are complicating factors to be taken into account (including transaction costs of collection of revenues at different levels and the interconnectedness of several of the benefits), the principles discussed above are reflected both in the way some industrialised countries finance sewerage investments and in the most innovative and appropriate forms of subsector financing observed in developing countries. In many communities in the USA, for example, households and commercial organisations pay for sewer connections, primary sewer networks are financed by a sewer levy charged to all property owners along the streets served, and secondary sewers and major collectors and interceptors are often financed by improvement levies on all property owners in the serviced areas.

Innovative sewerage financing schemes are now being observed in developing country cities. In Orangi, an informal urban settlement in Karachi, a hierarchical scheme for financing sewerage services has developed in which households pay the costs of their "on-lot" (*i.e.* on-site) services (*e.g.* latrines and septic tanks), the primary sewers are paid for by the households along the "lane" (public passageway between rows of houses), contiguous "lanes" pool their resources to pay for neighbourhood sewers, and the city (via the Municipal Development Authority) pays for trunk sewers.

The arrangements for financing condominial sewers by the urban poor in Brazil follow a remarkably similar pattern; households pay for the on-site costs, blocks pay for the block sewers (and decide what level of service they want from these), with the water company or municipality paying for the trunk sewers. Lack of access to credit may impede investment in sanitation, drainage and other essential urban environmental services, especially in small cities and towns. This problem has been overcome in some cases by creating special municipal development funds or rotating funds to finance environmental investments. For example, the World Bank has supported the creation of municipal development funds in the State of Minas Gerais, Brazil, for environmental improvements in small cities and towns, and in Mexico for municipal water supply, sewerage and solid waste investments in intermediate cities.

Similarly, poor urban households need mechanisms to finance sewer connections and in-home sanitary facilities. Some cities provide credit to poor households for these investments that can be paid off in installment payments (not subsidised) over periods of three to five years. Where there are well-managed water and sewerage utilities, the installment payments can be collected as part of the monthly water bill. In some cases, households can provide "sweat equity" (labour inputs provided by the community for self-help construction schemes) or even make partial payment in the form of construction materials.

A special sanitation credit fund has been established in Honduras for poor urban households, fashioned along the lines of the well-known Grameen rural credit bank in Bangladesh. Such experiences show that the urban poor will invest in a healthier environment if they can spread the initial costs over time. Similarly, innovative schemes for providing urban households access to credit for sanitation investments have been demonstrated in Lesotho and in Burkina Faso.

## WASTEWATER TREATMENT

Even when the appropriate financing and institutional principles are followed, very difficult issues can still arise with respect to the financing of wastewater treatment facilities. In industrial countries, two very different models are used. In many industrialised countries, the approach followed has been to set universal environmental standards and then to raise the funds necessary to finance the required investments. It is becoming increasingly evident that such an approach is proving to be very expensive and not financially feasible, even in the richest countries of the world. In the UK, the target date for compliance with the water quality standards of the European Union (EU) is being reviewed as customers' bills rise astronomically to pay the huge costs involved. In the USA, US$ 56,000 million in federal construction grants were provided to local governments from 1972-89 to build mandated secondary treatment facilities, but these grants have now been eliminated (and replaced by State revolving funds for loans to municipalities) at the same time that increasingly stringent environmental standards are being proposed.

Many local governments are now refusing to comply with the unfunded mandates of the Federal Government. The city of San Diego, for example, has refused to spend US$ 5,000 million on federally-mandated secondary treatment, arguing that it is more cost-effective to use long, coastal outfalls for sewage disposal. San Diego brought suit against the Federal Government and recently won its case in the federal courts. The US National Research Council has advocated a change in which costs and benefits are both taken into account in the management of sewage, with a shift to a water quality-based approach at the coastal zone, watershed or basin level.

In a few countries, a different model has been developed. In these countries, river basin institutions have been put into place which:

- Ensure broad participation in the setting of standards, and in making the trade-offs between cost and water quality.
- Ensure that available resources are spent on those investments which yield the highest environmental return.

- Use economic instruments to encourage users and polluters to reduce the adverse environmental impacts of their activities.

These institutional arrangements are described more fully below. In river basins in Germany and France, and more recently in Brazil, river basin financing and management models are applied in order to raise resources for wastewater treatment and water quality management from users and polluters in the basin. The stakeholders, including users and polluters as well as citizens' groups, are involved in deciding the level of resources to be raised and the consequent level of environmental quality they wish to "purchase". This system has proved to be efficient, robust and flexible in meeting the financing needs of the densely industrialised Ruhr Valley for 80 years, and for the whole of France since the early 1960s. There is growing evidence that if such participatory agencies were developed, people would be willing to pay substantial amounts for environmental improvement, even in developing countries.

In the state of Espirito Santo in Brazil, a household survey showed that families were willing to pay 1.4 times the cost of sewage collection systems, but 2.3 times the higher cost of a sewage collection and treatment system. In the Rio Doce Valley, an industrial basin of nearly three million people in south-east Brazil, a river basin authority is in the process of being developed. Stakeholders have indicated that they are willing to pay about US$ 1,000 million over a five-year period for environmental improvement.

In the Philippines, recent surveys show that households are often prepared to make substantial payments for investments which will improve the quality of nearby lakes and rivers. For developing countries, the implications of the experience of industrialised countries are clear. Even rich countries manage to treat only a part of their sewage, *e.g.* only 52 per cent of sewage is treated in France and only 66 per cent in Canada. As in the USA, Japan and France, most countries have provided some form of environmental grants to municipalities in order to achieve their present levels of treatment. Given the very low initial levels in developing countries (*e.g.* only about 2 per cent of wastewater was treated in Latin America at the beginning of the decade) and the vital importance of improving the quality of the aquatic environment, an approach is needed that simultaneously makes the best use of available resources and provides incentives to polluters to reduce the loads they impose on surface and groundwaters.

An effluent tax is one form of incentive that is used in many countries, ranging from France, Germany and The Netherlands to China and Mexico. It can be applied to any dischargers, cities or industries, with two benefits; it induces waste reduction and treatment and can provide a source of revenue for financing wastewater treatment investments. The dramatic impact of the Dutch effluent tax on industrial discharges is described by Jansen. The overall industrial effluent loads decreased by two-thirds between 1969, when an effluent tax was first applied, and 1985 (falling from 33 million to 11 million population equivalents).

The experience of China in the application of an industrial effluent tax for financing industrial wastewater management improvements has been described by Suzhen. In France and Mexico, the effluent tax is applied equally to municipal and industrial effluents, thus encouraging local investment in municipal wastewater treatment plants. An effluent tax, however, should be used in combination with municipal sewer use charges in order to ensure that industries do not escape paying for their discharges by passing the cost on to the municipality, as well as to ensure that the municipal sewerage authority has sufficient revenues to build and to operate sewerage and treatment works.

## COMMUNITY PARTICIPATION

The aspiration of most urban households, including the urban poor, is to have access to cost-effective and affordable sanitation services via public or private utilities. Consequently, they would be willing to participate, as responsible users, by paying the appropriate service charges. In the cities of many developing countries, however, such services are not yet universally accessible and poor communities must, themselves, get involved in the planning and delivery of sanitation and sewerage options.

The examples of the condominial sewer system in Brazil and the Orangi Pilot Project indicate an important institutional approach to community participation in which a productive partnership is formed between community groups and the municipal government or the utility. Often, such a system involves public provision of the external or trunk infrastructure, which may be operated by either the public or private sector, and the community providing and managing the internal or feeder infrastructure. The link between feeder and trunk infrastructure is essential for the evacuation and disposal of human waste collected by the community, but it is too easily overlooked. Many forms of community participation are possible for the provision of sanitation and sewerage services, such as:

- Information gathering on community conditions, needs and impact assessments.
- Articulation of, and advocacy for, local preferences and priorities.
- Consultations concerning programmes, projects and policies.
- Involvement in the selection and design of interventions.
- Contribution of "sweat equity" or management of project implementation.
- Information dissemination.
- Monitoring and evaluation of interventions.

Promoting and enabling community participation can take many forms. Where political will exists, governments may promote participation and create the conditions under which communities and households, as well as NGOs and the private sector, can play their appropriate roles. The World Bank-financed PROSANEAR project in Brazil for example, provides a framework and the resources for municipalities and utilities to experiment with innovative technical and institutional arrangements for providing sanitation services to the urban poor. When such government support is absent, alternative approaches have commonly been used to stimulate community involvement and to build the necessary political will.

First, NGOs or community-based organisations (CBOs) often play a catalytic role in mobilising communities and forming partnerships. In one of the largest scale examples involving an NGO, Sulabh Shauchalaya International began, in 1970, promoting the construction of pour-flush latrines in Delhi and other Indian cities, and over a period of 20 years assisted in building over 660,000 private latrines and 2,500 public toilet complexes with community participation and government support.

Second, consultations and town meetings are increasingly used as a forum to discuss and agree on environmental priorities, and to propose participatory solutions. Finally, communities may engage in public protests or legal actions as a means of building a constituency of the urban poor, and applying pressure on local governments and utilities for dialogue and action. The Orangi Pilot Project had its origins in the discontent of local residents with excreta and wastewater overflowing in the streets as a result of the failure of the Karachi Development Authority to provide adequate sewerage.

## A ROLE FOR THE PRIVATE SECTOR

Financial resources can also be mobilised through the private sector; poor service provision by the public sector often suggests a need for increasing partnerships with the private sector. Private sector participation, however, is only one possible opportunity; it is not a panacea.

In situations in which existing sanitation service delivery is either too costly or inadequate, private sector participation should be examined as a means of enhancing efficiency and lowering costs, and of expanding the resources available for service delivery. In deciding whether to involve the private sector, it is important to assess several key factors which have been summarised by the *Infrastructure for Development: World Development Report, 1994*. Introducing competition is the most important step in creating conditions for greater efficiency by both private and public operators; some services can be split into separate operations to help create contestable markets. The principle of accountability to the public should be maintained through transparent contractual agreements that are open to public scrutiny and should help to minimise risks to public welfare, create real competition, ensure efficiency, and promote self-financing.

Paradoxically, public sector capacity may have to be strengthened in order to achieve effective private sector participation which requires public sector agencies with sufficient capacity to prepare bidding documents and performance indicators, assess proposed outputs and costs, administer the contracting process, and regulate contract performance. In Mexico, municipalities are granting concessions to the private sector to build and operate wastewater treatment plants, both as a means of financing investments in plants through the private sector and to overcome problems with weak local operating capacity.

The Puerto Vallarta wastewater treatment plant was the first of many new plants to come on line in the past few years. An important point to remember in cases such as Puerto Vallarta is that the private sector performs the necessary function of mobilising financing for needed investments, but the investments made together with operations, maintenance and depreciation costs will all have to be recovered through tariffs charged to domestic and industrial customers. Another innovative example is a concession to 26 industries in the Vallejo area of Mexico City to form a new enterprise, Aguas Industriales del Vallejo, to rehabilitate and expand with its own funds an old municipal wastewater treatment plant, treat up to 200l s-1 of sewage, and sell the treated water to shareholders at 75 per cent of the public utility water tariff.

## STRATEGIC PLANNING AND POLICIES FOR SUSTAINABLE SANITATION SERVICES

Applying a strategic planning approach to urban sanitation problems should result in choosing the right policy instruments, agreeing priorities, selecting appropriate standards for service provision, and developing strategic investment and cost recovery programmes. The question of appropriate service standards is a particularly vexing one that, in the end, should be answered by considering user preferences and willingness-topay. In a large city with many pockets of poverty, service standards are likely to be spatially differentiated because many households cannot afford conventional sewerage without massive government subsidies. The Kumasi Strategic Sanitation Plan provides an example of a differentiated plan matching housing types, income levels and user preference; the plan recommends that sewers be used in tenement areas, latrines in the indigenous areas, and flush toilet/septic tank systems in high income and new government areas. Willingness-to-pay surveys were carried out, and the results were used to help define differentiated financing options.

Explicit subsidies were targeted to the city's low-income population. Municipal wastewater treatment is a particularly costly and long-term undertaking so that sound strategic planning and policies for treatment are of special importance. The recently endorsed Environmental Action Programme for Central and Eastern Europe (CEE), formulated with the assistance of the World Bank, recognises that the CEE countries will require a plan to move towards Western European standards over a period of 15-25 years as financial resources become available. Although urban sewerage levels in the CEE are generally adequate, 40 per cent of the population are not, at present, served by wastewater treatment plants. The domestic pollution load represents 60-80 per cent of the combined municipal and industrial organic waste load in many CEE cities. Furthermore, many of the existing plants are currently overloaded, poorly operated and maintained, or bypassed.

The following is a checklist of policy questions posed in the CEE Action Programme to be answered before proceeding with municipal waste-water investments:

- Have measures been taken to reduce domestic and industrial water consumption?
- Has industrial wastewater been pre-treated?
- Is it possible to reuse or recycle wastewater?
- Can the proposed investment be analysed in a river basin context? If so, have the merits of the investment been compared with the benefits from different kinds of investments in other parts of the river basin? (Note that a least-cost solution to achieve improved water quality may involve different, or no, treatment at different locations.)
- Has the most cost-effective treatment option been used to achieve the desired ambient water quality?
- Has there been an economic analysis to assess the benefits (in terms of ambient water quality) that could be achieved by phasing investments over 10 years or more?

## COST-EFFECTIVE TECHNOLOGIES

Developing country cities are beginning to recognise that poor urban residents cannot afford, nor do they necessarily want or need, costly conventional sewerage. Beyond the dense urban centres, the average household cost of conventional sewerage may range from US$ 300-1,000. This is clearly too expensive for many households with annual incomes well below US$ 300. Fortunately, a broad range of cost-effective technological options are available to respond to the demands of urban consumers beyond the urban centre, with the potential to reduce costs to the order of US$ 100 per household.

The UNDP/World Bank, Water and Sanitation Programme has worked with many countries over the past decade to develop, demonstrate, document and replicate many of these lowcost sanitation options. The examples drawn upon throughout this chapter illustrate many of the options available to households (*e.g.* ventilated improved pit (VIP) latrines in Lesotho, Sulabh pour-flush latrines in India, condominial sewers in Brazil and simplified sewerage in Pakistan), as well as the supporting institutional and financial systems that make possible the wide-scale application of these options. Wastewater treatment technologies also have a wide range of costs. Conventional treatment processes may cost US$ 0.25-0.50 per cubic metre. If non-conventional options can be used, it may be possible to cut these costs by at least onehalf.

Promising low-cost treatment approaches, especially for small and intermediate cities, range from natural treatment systems (such as waste stabilisation ponds, engineered wetlands systems and even ocean outfalls), to decentralised treatment systems (such as are used in Curitiba, Brazil), to new treatment processes (for example anaerobic treatment processes such as the upflow anaerobic sludge blanket (UASB) reactors presently operating in cities in India, Colombia and Brazil).

In large cities, land or other constraints may result in conventional treatment being the most cost-effective approach for achieving the desired water quality objectives, although this should always be a decision resulting from an economic analysis. Lifetime costing should always be used to compare and to choose among treatment options, because operations and maintenance constitute a major share of the costs.

## CONSERVATION AND REUSE OF SCARCE RESOURCES

Cornerstone ecological principles for sustainable cities include the conservation of resources and the minimisation and recycling of wastes. Translating these principles into urban policies for wastewater management should emphasise the strategic importance of water conservation and wastewater reclamation and reuse in cities.

Successful conservation and reuse policies, moreover, need to achieve a balance between ecological, public health and economic and financial concerns. Pricing and demand management are important instruments for encouraging efficient domestic and industrial water-use practices and for reducing wastewater volumes and loads.

Water and sewerage fees can induce urban organisations to adopt water-saving technologies, including water recycling and reuse systems, and to minimise or eliminate waste products that would otherwise end up in the effluent stream.

In addition to pricebased incentives, demand management programmes should include educational and technical components, such as water conservation campaigns, advice to consumers, and promotion, distribution or sale of water-saving devices like "six-litre" toilets which use less than half the volume of water per flush than a standard toilet.

Wastewater reclamation and reuse is increasingly recognised as a water resources management and environmental protection strategy, especially in arid and semi-arid regions. The use of reclaimed urban waste-water for non-potable purposes, such as in-city landscape irrigation and industry or for peri-urban agriculture and aquaculture, offers a new and reliable resource that can be substituted for existing freshwater sources. Water pollution control efforts can make available treated effluents that can be an economical source of water supply when compared with the increasing expense of developing new sources of water.

Conversely, in developing countries only recently embarking on major wastewater treatment investments, reuse has the potential to reduce the cost to municipalities of wastewater disposal. A framework for the economic and financial analysis of reuse projects has been provided by Khouri *et al.* in a planning guide that integrates economic, environmental and health concerns with agronomic concerns for the sound management of crops, soil and water.

# Bibliography

A.J. Macself: *Soils and Fertilizers*, Satish Serial Publication, Delhi, 2005.

D.R. Khanna, R. Bhutiani and Gagan Matta: *Water and Wastewater Management Volumes I-II*, Daya Publication, Delhi, 2010.

David Norman and Mancolm Douglas: *Farming Systems Development and Soil Conservation*, Daya Publication, Delhi, 2007.

Deo Kant Prasad: *Soil Conservation and Organic Farming*, Enkay Publication, Delhi, 2012.

E. Jackson: *Crop Management and Soil Conservation*, Biotech Books, Delhi, 2011.

E. Ramann. Translated by C.L. Whittles: *The Evolution and Classification of Soils*, Asiatic Publication, Delhi, 2006.

Gilbert Wooding Robinson: *Soils : Their Origin Constitution and Classification*, Biotech Books, Delhi, 2005.

Helmut Kohnke and Anson R. Bertrand: *Soil Conservation*, Biotech Books, Delhi, 2009.

Herminie Broedel Kitchen: *Soils and Crops : Diagnostic Techniques*, Satish Serial Publishing, Delhi, 2004.

Hough Hammond Bennett: *Soil Conservation for Sustainable Agriculture*, Agrobios Publication, Delhi, 2001.

Hugh Hammond Bennett: *Elements of Soil Conservation*, Biotech Books, Delhi, 2009.

John Atkinson: *The Mechanics of Soils and Foundations*, Taylor and Francis Group, 2010.

Klein Gomes: *Wastewater Management*, Oxford Book Company, Delhi, 2009.

L L Somani and P C Kanthaliya: *Soils and Fertilisers at a Glance*, Agrotech Publication, Delhi, 2004.

Madireddi V. Subba Rao: *Soil Conservation Management and Analysis*, Daya Publication, Delhi, 2014.

Madireddi V. Subba Rao: *Water Conservation Management And Analysis*, Readworthy Publication, Delhi, 2011.

Mahendra Pandey: *Wastewater Management*, Dominant Publication, Delhi, 2011.

O.P. Gupta: *Water in Relation to Soils and Plants : With Special Reference to Agriculture*, Agrobios Publication, Delhi, 2002.

P P Mahendran; P Saravana Pandian; P Balasubramaniam and A Saravanan: *Soil Resource Inventory and Management of Problematic Soils*, Agrotech Publication, Delhi, 2008.

Paul J. Kramer and John S. Boyer: *Water Relations of Plants and Soils*, Academic Press an imprint of Elsevier, 2013.

Pietro Laureano: *Water Conservation Techniques in Traditional Human Settlements*, Copal Publishing Group, Delhi, 2013.

S C Panda: *Soil Water Conservation and Dry Farming*, Agrobios Publication, Delhi, 2007.

S K Datta: *Soil Conservation and Land Management*, International Book Distributors, Delhi, 2006.

S S Negi: *Hand Book of Soil Conservation and Integrated Watershed Development*, International Book Distributors, Delhi, 2007.

S.C. Panda: *Soil Conservation and Fertility Management*, Agrobios Publication, Delhi, 2008.

S.K. Saini and Subhash Chandra: *Water Conservation and Utilization in Agriculture Production*, Satish Serial Publishing House, Delhi, 2010.

S.N. Kaul, Tapas Nandy, L. Szpyrkowicz, A. Gautam and D.R. Khanna: *Wastewater Management : With Special Reference to Tanneries*, Discovery Publishing House, Delhi, 2005.

S.N. Kaul: *Wastewater Management in Cluster of Small Scale Industries*, Book Enclave, Delhi, 2001.

T.V.S. Prasad: *Soil Chemistry : Nutrient and Water Management in Agricultural Soils*, Dominant Publication, Delhi, 2009.

V.V. Dhruva Narayana: *Soil and Water Conservation Research in India*, Indian Council of Agricultural Research, 2002.

# Index

### F



### G



### H



### I



### K



### L



### M



### N



### O



### P



### R



### S



### T



### V